Eve's Green Fingers

A Cultural History of Plants

Martin Ingrouille

Fat
Dog
Books

Copyright/Disclaimer

Eve's Green Fingers by Martin Ingrouille

Contents

But lo! men have become the tools of their tools. The man who independently plucked the fruits when he was hungry is become a farmer; and he stood under a tree for shelter, a housekeeper.

Henry David Thoreau 1817-1862

This book is a short history of the place of plants in human culture and of the relationship between plants and humans.

1 Eve's Green Fingers

The opening sequence in the film 2001 Space Odyssey imagines the moment apes became human. A mysterious black obelisk makes the ape figures pick up bones and use them as tools. In the moment they use them as weapons, flinging them into the sky, the apes are transformed and become human. It is a powerful image; the bone/tool is transformed into the spaceship traveling in a beautiful arc towards the space station to the music of the Blue Danube. The Biblical image is different but closer to the truth. God, the gardener breathes life into the soil to create Adam.

"And the LORD God formed man of the dust of the ground, and breathed into his nostrils the breath of life; and man became a living soul." Genesis 3 7

But rather imagine the key creation moment like this. To the music of Debussy: a woman dressed in skins is digging up a taro or yam. She is a female, because women normally play the key role of the primary gatherer in hunting-foraging societies. The woman pauses and thinks, she takes the tubers but pulling the leafy crown off one she replants it and, with short grubby fingers, firms the soil back around it. In that moment, Eve lit a green fuse, and human civilization was born.

At the birth of taxonomy, the science of naming things, a time when it was believed that several species of humans existed, the Swedish taxonomist Linnaeus pondered hard about what to name civilised humans, eventually settling on what has proved to be extraordinarily ironic, *Homo sapiens*, from sapiens "wise or judicious". Another human species he named was *Homo troglodytes*, for the cave-living species of human that had been reported from various parts of the world; the "troglodytes" part of the name was transferred to chimpanzees *Pan troglodytes*, though they do not live in caves. How much better if he had named humans *Homo cultor*, "man the cultivator". The thing that characterizes humans above all other animal species is that they are they are cultivators and gardeners. It was the betrothal of plants and humanity that was to transform the world. The Latin word itself, "cultor", as well as meaning cultivator, planter or gardener, has other highly significant subsidiary meanings, such as "resident or inhabitant" and "worshipper", recognizing the primal link between plant cultivation, and a settled life-style and spiritual culture that accompanies it. It's a link that is more firmly established yet by our uses of the derivative word "culture", to mean not just to grow but also for the products of that growth, including all the things that make us distinctly human, our culture.

Apes and Human Diets

Human-apes and apes, and animals of all sorts have been using plants for ever, because without plants there would be no life on land beyond stripes of yellow and purple and green slime in damp hollows. And humans are not the only gardeners. There are ants that plant seeds. There are birds that adorn their nests with flowers. There are bees that spice up their sex life by collecting perfume from flowers.

Chimpanzees are our closest living relatives. They eat a variety of plants, mainly ripe fruits, but also foliage, and they fashion tools from plants, when for example, for a treat, they winkle termites out of their nest with a twig or sprig of bamboo. They crack nuts open between two hard stones. One study of Chimpanzees records the use of 120 different species of plant. They exploit some fruits like figs which are produced, though by different trees, over much of the year, as a staple, and add diversity to their diet as other fruits become available. Some plants they utilize not for food but for health or to build nests. They swallow whole leaves of *Vernonia amygdalina*, a tree with clusters of small white daisy like flowers, to control intestinal nematode infections. They seek out a plant with yellow marigold like flowers, called *Aspilia latifolia*, if they have a stomach problem; it contains an anti-bacterial compound called thiarubrine-A. Humans have also learned to use it for stomachaches and wounds, and have also learned that many other plants in the same daisy family Asteraceae have medicinal or insecticidal properties.

Paleolithic diets were based on wild game and fish and uncultivated plant foods, with carbohydrates from root crops and fruits but not from cereals, and sugar only from fruit and some honey. The earliest humans were probably at least as sophisticated in their diet as modern chimpanzees, though, unlike apes, quantitative analyses of human hunter-gatherer diets indicate that a greater proportion of their diet, on average 65% of their energy, came from animal sources and only 35% from plants. In time humans relied more and more on plants and less on animals, rather like the great apes today, but with a vital difference, that they developed a kind of gardening of nature. In the same way surviving human hunter-gatherer communities protect their environment even while they exploit it, by actively propagating wild plants.

Horticulturalists

"The Lord God took man, and put him into the Garden of Eden to tend and keep it dress it and keep it. Of every tree of the garden you may freely eat." Genesis 2.16,17

The transition from a hunter-gatherer type subsistence to a full menu of vegie-culture gardening was not a difficult one to make. There are still surviving cultures who live by vegie-culture in many parts of the tropics. Unfortunately, vegie-culture has not left much of an archaeological record so that it is difficult to determine whether the earliest humans were just exploiting wild plants or actively propagating favored ones. To get a fuller picture of early plant use by humans, we must turn to historical records of what hunter-gatherer communities were doing, when they were first encountered by agricultural humans, for example at the time of first European contact in 1789 the aboriginal people of the Sydney area. The largest group, called the Darug, cultivated yams (*Dioscorea*) and lilies with edible tubers (*Arthropodium milleflorum*) as their main root crops. "Darug" is the aboriginal name of the yam. These people were soon largely exterminated by small-pox but anyway within a few years of European colonization largely destroyed the native culture when European farmers took over the river flatlands where they had cultivated their yams.

The Darug divided plant food into three categories: food that was dug up like the yam – *burrawang*, fruits and seeds - *wigi*, and nectar – *watangal*, sucked directly from flowers or shaken with water to make a sweet drink called *bool*. As well as yam roots and lily tubers, rhizomes were obtained from bracken (*Pteridium esculentum*).

Fruits came from species such as the lillypilly (*Syzygium paniculatum*), *Exocarpos*, blue-flax lily (*Dianella caerulea*) and figs (*Ficus*). An important food was burrawang, the seeds of the cycad *Macrozamia communis*. The use of these illustrates an important aspect of human plant use that contrasts with other animals – food technology or cooking. The seeds of *Macrozamia* were first treated to remove toxins by pounding and soaking in water for at least a week and then roasted as cakes. Watangal came from species such as *Banksia ericifolia*, the grass-trees *Xanthorrhoea resinifera* and waratah (*Telopea speciosissima*). The range of plants used is much greater than the few mentioned here so that no one species was over-exploited. In addition, they husbanded these resources in various ways like breaking off the top of the tuber after digging out the yam and replanting it to re-establish the plant, by discarding seeds around middens where they would germinate, and by not collecting some fruits after they had fallen to the ground so that the fruit trees could regenerate.

The essence of the vegie-culture in the tropics lies in the diversity of plants used and the close relationship of the forest to the garden. These gardens truly represent the Garden of Eden. The Kwara'ae tribe of the Solomon Islands describe the forest:

"The forest is for living in and for developments such as gardening and home-building in, and also for taking wild growing things and living things to eat and work with"

There is a widespread belief in nature spirits, but it is not airy-fairy nonsense but deeply rooted in a knowledge of the plants. They do not seem to have plant gods that are objects of cult and the use of plants is not normally as ritualized as the cultivation of cereals was to become. Rather their respect is born from knowledge of what they are useful for and how to use them, their use is governed by the petty rituals of hearth home and garden - "That's how we do this here". "This is what we use for that here".

Some plants are not for food or craft but are planted as protection plants, they keep pests off a favored crop, or because they are pretty, such as a line of *Coleus* around the houses or in gardens to protect taro and yams. Not all plants are useful, *"This is just a weed, it is not good for anything"*, but most plants have some use, *"this is a wild vegetable*", "*we cure with it*", "*when small we build houses with it*".

"Sweeter" roots

It is the cultivation of root crops that provides the foundation for this kind of life. Tropical root crops are usually derived from climbers and scrambling plants that grow in rather more open and marginal situations in the forest such as the wetter soils on riverbanks. Instead of producing a freestanding woody trunk they store up energy in basal organs such as tubers and rhizomes from which repeatedly they put out fast-growing stems that twine up other plants. Although the three major lowland tropical root crops are from three different families: cassava (Euphorbiaceae), sweet potato (Convolvulaceae) and yams (Dioscoreaceae), they share many features. Frequently they have heart-shaped or lobed leaves. Of great significance to humans, the storage organs are often protected from bacterial and fungal rot, and nematode and insect attack, by the presence of toxic compounds. Cassava roots produce cyanide, and taro and yams are rich in calcium oxalate crystals that can cause the throat to swell up and close. Perhaps the first step in the development of vegie-culture was not the propagation of the plants themselves but by the discovery that peeling and cooking the tubers or rhizomes would remove or destroy many of the toxic compounds present. So it is the use of fire for cooking that separates human-apes from other apes. In time sophisticated methods of processing the food was developed.

In the process of domestication "sweeter" low toxin variants of root vegetables were selected. Paradoxically as humans became more dependent on these low toxin variants the more the plants themselves relied on humans to protect them from destruction and propagate them. A contract had been signed and the human and the crop were tied into co-dependency. With their storage organs harvested they had less energy to produce flowers and seeds and required humans to vegetative propagate them. Humans noticed that if they were cultivated in the same spot too long the yield went down, partly because of a loss of soil fertility and partly because of a build-up of pests and diseases, exacerbated if the plants lacked the toxins that would otherwise have protected them. This meant moving to a newly cleared patch of ground whenever an established garden had become too infested with disease, and the development of a kind of shifting cultivation that was almost universal in the tropics until recently.

The cultivation of yams predates by millennia the cultivation of cereals and is still widespread in the tropics. The twining stems are either given artificial support or grown up trees. Different yams have been taken into gardens and domesticated in different parts of the world: in Asia *Dioscorea alata*, in Africa *Dioscorea rotundata/cayenensis* and in America *Dioscorea trifida*. They share twining stems with alternately arranged heart–shaped leaves but differ in the form of their basal tubers, either single or multiple, and lobed or un-lobed. *D. alata* sometimes produces a single large lobed tuber that can weigh up to 20kg. It has an Asiatic origin but has become very widespread, though in Africa the yellow yam *D. cayenensis* is preferred because of its higher starch content. *D. bulbifera*, the potato yam produces aerial tubers from the places where the leaves are attached to the stem, as well as underground tubers. Only the aerial tubers are edible. In the period of European exploration yams were favored as a ship's victual because they also have a long shelf-life unlike cassava, and they are rich in vitamin C and combated scurvy. They became widely distributed along trade roots. Their long shelf-life may be due to the presence of dioscorins, a class of storage proteins that have biological activity, some acting as enzyme inhibitors or lectins (compounds that bind to specific carbohydrates) which may act against rot causing bacteria and fungi. The dioscorins are toxic to humans too but are destroyed when the tuber is cooked.

The cultivation of root crops dates in the Solomon Islands dates at least as far back as 28,000 years. The account of the present day Kwara'ae Solomon Islanders provides an insight to into a sophisticated gardening of nature that is probably many millennia old. The Kwara'ae eat a variety of *Dioscorea* species both from the wild and their gardens (*D. alata*, *D. esculenta* and *D. bulbifera* among others) and recognize many different varieties of these. For example, Kamo is wild *D. alata*, Kwaru'abu, Kwarubalu, Ludaboasi, Luluka, Idubiru and Borosura are varieties of gardening yam differing in the shape of the tuber, the color of their flesh and their pattern of growth. Mounded or Pluckhead Gwa'ufi is a wild yam.

"When we dig it up to eat we pluck off the tuber at the base, then we mound over the base and heap ground on the site of the tuber for the time when we will dig it again. Or we dig and lift some tubers and some tubers remain if we pluck off the tuber."

However traditionally taro, also called dasheen or cocoyam, *Colocasia esculenta* is the most important root plant to the Kwara'ae, and again there are numerous varieties. It produces an erect root stock scarred where leaves have fallen, weighing up to 4kg. It prefers swamp conditions. Today it is part of the daily diet of at least 100 million people. It has been domesticated for millennia, and now rarely flowers, relying on humans to propagate it. Like other tropical root crops this is relatively easy to do by just replanting the crown of the plant after taking the corm. The ease and naturalness of the transition from collecting from wild plants, through the propagating those wild plants by replanting part of the plant after harvesting it, to their cultivation in gardens is shown by the diffuse origins of this kind of plant use. In some cases, it was the same or a different but closely related species that was separately domesticated in different places. So for example in the New Guinea Highlands taro and yams and bananas were separately domesticated by about 7000 B.C.E. In the Kuk basin forest plots were being cleared and elaborate systems of ditches and diverted water-courses established. The picture is complicated because especially following western expansion these asexually propagated crops were widely distributed and introduced to new areas.

Colocasia has an origin in south-east Asia but there is a plant in the same family, the Araceae, called tannia or yautia, also sometimes cocoyam, *Xanthosoma sagittifolium* that was domesticated in the American tropics. Like many of these tropical root crops it has unpalatable constituents. In tannia there are calcium oxalate crystals and saponins that have to be removed, in this case by peeling and roasting. Imported into West Africa tannia has displaced taro and even yam as the primary root crop. South America has been a rich source of root crops. They include cassava, arrowroot, sweet potato, yampee, achira, jicama, oca, ullaco, ysano, maco, arracacha and of course the king of vegetables the potato.

In the lowland tropical Americas, it was cassava that was perhaps the most important plant to be domesticated in this way. It is a staple food for 500 million people today. Cassava or manioc *Manihot esculenta* (family Euphorbiaceae) is the most important tropical root crop and provides more than half of the energy requirement for over 420 million people in tropical countries. Its root is almost pure starch. It is an herbaceous to semi-woody perennial that can grow several meters tall and has long-stalked palmate leaves. It has the advantage over many other root crops that it is relatively drought resistant. There are two forms distinguished by the content and distribution of hydrocyanic acid (HCN) in the tubers. Sweet cassava has low HCN with HCN confined to the tuber bark. Bitter cassava has high HCN distributed throughout the tuber. The HCN is stored as a sugar, a cyanogenic glycoside called linamarin, and is released by the action of enzymes when the plant is damaged. Cooking destroys the cyanogenic glycoside. Sweet cassava (aipim, macaxeira) was cultivated, and perhaps, first domesticated by the Mayans. It is more fibrous but can be peeled and boiled like potatoes. Bitter cassava may have an origin further south in Tropical America.

Cassava was introduced into West Africa by the end of the 16th century that is within a hundred years of Columbus' "discovery" of America. In many parts of West and Central Africa it is pounded into a kind of mash called fufu. Yam and tannia and plantain are also used to make *fufu*. American Indians use a kind of elongated basket woven from palm fibers to extract the starch. Small pieces of cassava are grated into the tube called a tipitis which is then kneaded so that the starch collects at its base. Fresh cassava has a very limited shelf-life but is readily converted into flour. Peeled tubers are grated, dried and then ground up to produce a flour-like *farinha de mandioca*. Tapioca is produced by kneading grated tubers in water to remove the starchy milky juice, and then washing and drying the starch that remains; heating on a flat plate causes the starch grains to pop and stick together. The juice can be boiled down to destroy the cyanide and make a sauce called tucupi. The flour is used in various ways: in Brazil there are sun dried cakes called carimã, crisps called beijus or it is mixed with ground fish to make paçoka.

Sweet potato

Sweet potato or kumara, *Ipomoea batatas* (Convolvulaceae), also has a Central or South American origin. Fossilized sweet potatoes have been discovered in the Andes from 8,000-10,000 years BC. The "sweetness" refers to the lack of toxic constituents so that kumara is relatively easy to use and now it is widely available even in temperate regions and roasted, mashed or made into a delicious soup. Wild relatives contain the glucoside ipomoein that is a purgative, so the sweetness of kumara may be the result of early domestication. There are hundreds of varieties, with white, orange or purple flesh, and differing in the sweetness and smoothness of the flesh. Kumara was one of the crops Columbus brought back to Europe and from there was introduced throughout the Old World by the Portuguese and Spanish. The Kwarae'ae record its introduction from Fiji and Queensland by their ancestors, perhaps indicating a fairly recent introduction.

However, there is archaeological evidence, of an earlier introduction of sweet potato to Polynesia. The Chinese found sweet potato in the Philippines when a governor of the Fujian province sent an expedition there. Its origin in Polynesia is mysterious but there is evidence, from microfossils, for prehistoric cultivation in Aotearoa (New Zealand), probably the last place it was introduced, though this may not have been until 1100 C.E. or later. It was kumara that permitted the colonization of New Zealand. The colonists also brought breadfruit and cocoanuts but they did not flourish in the cooler New Zealand climate like kumara. They cultivated it in mounds, along with bracken (*Pteridium esculentum*) for its rhizomes, and although it is a perennial, they treated it like an annual by storing the tubers through winter in special rimmed storage pits. It was one of the staples by the time of extensive European contact, although new varieties had by then been introduced by whalers and sealers, following Abel Tasman's first western contact in 1642.

There is linguistic evidence for an earlier South American origin because Sweet Potato is called cumara in Peru and kumara in Oceania. Carbon dated remains of sweet potato have been found in Peru dating from between 8-10 thousand years ago and the earliest representation of sweet potato on pottery are also from Peru. However if kumara was introduced by human contact it is unlikely to have been from Peru, but rather from Central America; molecular variation indicates that Mexican kumara is more closely related to that in Oceania, and more variable than that in Peru. Thor Heyerdahl showed that an introduction to Polynesia from the west was possible by his voyage in the balsa (*Ochroma lagopus*) raft Kon-Tiki. Certainly at the time of their colonization of the outer reaches of Oceania the Polynesians were perfectly capable of reaching the Americas in their multiple outriggers that rivaled the size of contemporary Chinese and European vessels. It was the development of these vessels that after 300 BC allowed the westward expansion of the Polynesians from Samoa and Tonga. They reached Hawaii by 600 C.E. and the Society Islands and Tahiti by 800 C.E. and later Easter Island. It is strange though that if the Polynesians reached the Americas at around this time that it was kumara alone that made this journey back to Polynesia and not the myriad of other crops, including maize, that were then available. Perhaps the spread of kumara into Oceania was by earlier natural long distance dispersal.

Another plant introduced by the Maoris was a vine, the bottle gourd or calabash plant (*Lagenaria siceraria*). This has certainly spread through long-distance dispersal; the gourds float easily. The genus is centered on Africa but *L. siceraria* has a pantropical distribution. There are different subspecies in South America and Africa. The flesh of the young fruit is edible, boiled, stir-fried, roasted or made into soup, but the older fruit is hollowed out and dried to make containers of all sorts. In China the bottle gourd Woo Lo Gwa is revered. "Woo Lo" describes its feminine wasted shape. Sages and holy people are frequently portrayed with a bottle-gourd or two hanging from their belt.

Other root crops

A plant that has traveled to Asia from the Americas is *Pachyrrhizus* Yam bean. There are several species with a South American origin that were carried west by the Spanish. It is a legume that produces a tuber with a lobed bottom. *P. erosus* or Jicama is Central American, but was introduced to the Philippines and from there it has spread widely in Polynesia and as far as southern China, Singapore and India. In the Philippines it is called sinkamas and in China, saa got. *P. tuberosus* from Ecuador is called the Chop-sui Potato in Hawaii. *P. ahipa*, aricoma, is a non-climber cultivated from Bolivia and northern Argentina. The pod and beans are poisonous but the tubers have a sweet and crunchy flesh.

Another legume vine with an edible tuber that also has an ancient use is Kudzu or Fun Got, *Pueria thunbergiana*. It can produce an elongated tuber more than 50cm long with a tough white sweet flesh that is 25% starch. Nowadays it is mainly processed to produce a thickener for soups and sauces but before the introduction of sweet potato it had a more general use as a staple food in South East Asia, grated into soups and stews. Unfortunately older and larger tubers are rather chewy. Nevertheless Kudzu has had its revenge on the Americas. It was introduced to the USA in 1876 from Japan to be grown in a Japanese garden at a centennial exposition in Philadelphia. Its large leaves, and sweetly smelling flowers made it a hit and it became widely grown. From the 1930s it was widely planted by FD Roosevelt's Conservation Corp in a misguided attempt to prevent soil erosion in the southeastern states. Farmers were even paid by the Soil Conservation Service to plant it. It now covers 7 million acres and is widely regarded as a highly damaging weed because of the way its rapid growth smothers other plants.

All these root crops were very important. Their domestication and cultivation of root crops probably predates that of cereals by millennia but it did not lead to the development of civilizations. It was a sustainable pattern of land use; in the lowland tropics of the Amazon and Orinoco basins a pattern of land use based on the cultivation of cassava and sometimes sweet potato, arrowroot and peanut was established about 1500 B.C.E. and has been sustained to this day. However it did not provide a surplus of storable food, material wealth. It required Adam and Eve to be expelled from the Garden of Eden for human culture to develop into the patterns we see today and crucially it was the cultivation of cereals that led to the development of complex civilization.

A thousand uses: palms and bamboo

In tropical horticulture many species have multiple uses. Above and around the vegetable garden the trees, especially the palms are notable for this. The Kwara'ae of the Solomon Islands use the cocoanut palm (*Cocos nucifera*) "milk" for drink, the flesh for cocoanut pudding and cream, make cups with the shell, the trunk for platforms, and for carving, the fronds are used for shelter and plaiting and weaving into baskets and trays, to make bats and balls for toys, and to wrap food, the leaf-spines are used for brush bristles, the leaf cuticle for plaiting armbands, the fruit coat for combs and decorations, the soft roots for dye and the skin for tea and medication. They don't just use coconut palms but several other species for a variety of purposes. There are nearly 3000 species of palm many of which provide drink, seed, fruit, timber, and fiber. Almost all parts of palms are utilized by humanity. The Borassus palm (*Borassus flabellifer*) has 801 uses according to a Tamil song: it provides timber, leaves for thatch and making baskets and mats, fibers for paper, seedlings are ground for flour, fruits roasted to eat and the inflorescence tapped to make palm wine. Palm wine, is produced by tapping the immature male inflorescence several other species including *Elaeis guineensis* and *Nypa fruiticans*, the mangrove palm.

In the American tropics, for people like the Yanomamo, the peach palm, pejibay or chonta, *Bactris gasipaes*, is a kind of staple, not least because it has multiple uses. Most importantly it produces large bunches of fruit, of about 100 fruits, weighing together up to 15 kg. The trunk is covered in spines but the forest people clear a space next to each tree and plant a fast growing *Cecropia* tree next to it. They climb up this and hook off bunches of fruit from the palm. When cooked the fruit flesh provides a staple for many forest people. By chewing and adding saliva a kind of alcoholic porridge is produced. It can even be dried and made into a kind of flour. The almond shaped seed inside tastes of cocoanut and is oil rich. The flowers are a salad vegetable. The wood is used for bows and spears, blow-pipes and darts, or poles. The leaves provide thatch. Some kinds produce multiple stems and so the tree can be harvested for palm hearts or palmito, the growing bud, without killing it the whole plant, and so it is one of the most widely cultivated palms for this purpose.

Palms grow not just in the wet tropics but also in drier areas. One of these drier land specialists is perhaps the most important palm fruit tree. Date palms, *Phoenix dactylifera*, provide large quantities of fruit with yields often in excess of 30 kg per tree. There are several varieties perhaps formed by hybridization between *P. dactylifera* and other species. Favored trees have been propagated from vegetative suckers for millennia. Date palm is dioecious and wind-pollinated but it was early understood that the amount of fruit set could be enhanced by pollination by hand. Today in date palm plantations a ratio of 1 to 25 or less male to female trees is maintained to provide pollen. An interesting phenomenon is that the male pollen directly affects the quality of the developing maternal fruit tissues (metaxenia).

Many different species provide oils. Oil palm, *Elaeis guineensis*, is one of the most productive of all crops. In a mature plantation of 8 to 20 years of age, good management produces up to 30 t/ha. Bunches of fruits weigh about 25 kg. The fleshy fruit has up to 70% oil content. Palm oil is used in wide variety of foods and in the manufacture of margarines, soap and cosmetics. The center of origin of oil palm is in the wetter parts of West Africa but there is now a very substantial plantation based industry established in South East Asia. Palm oil has a markedly different fatty acid spectrum, 44% palmitic acid a C16 saturated fatty acid but with a roughly equal content of unsaturated fatty acids. Other oil palms are exploited on a more limited scale, such as the Babassu palm *Orbignya phalerata* in Brazil.

The stem of the palm is utilized in several ways. The sago palm, *Metroxylon sagu*, and moriche, *Mauritia flexuosa*, provide starch from their pith. Palm hearts are the center of the apical buds. The leaves are used for thatch and weaving. Rattan palms are climbing palms that provide the long flexible stems extensively used in furniture. Many species also provide fiber or timber from the trunk. Fuel comes from all parts of the plant. Palms are also extraordinarily beautiful plants, widely planted for their ornamental value. Many tropical gardens are adorned with stately avenues of palms such as the royal palm.

Cultivated palms are identical or little different from wild plants though they may nowadays they may be planted in plantations. For example only in 1964 was the Oil Palm Institute OPRI established to produce improved, better cropping and disease resistant cultivars. It has been very successful, not so much in producing disease resistant varieties, but in promoting the growing of oil palms and now huge areas of natural forest in South East Asia and elsewhere are being converted into oil palm plantations. The Garden of Eden is in retreat and everywhere the age old traditional patterns of vegetable gardening are under threat. Nevertheless surviving tropical gardening brings us closest to that moment when Eve first gardened. The tropical forest and the garden plots it contains are the garden of Gardens vary from the clearing of a space around favored wild growing plants through various degrees of clearing, fencing and planting. Barren trees are encouraged to fruit by partially tripping the bar at the base or are pruned. Protector plants such as *Coleus* and *Hediotis* are planted to protect the crop plant. Bark is collected and scattered to increase fertility or protect from disease. Eventually gardens are overcome by weeds and are abandoned. In five to fifteen years tall trees have grown up again shading the weeds out and the patch is suitable for gardening again. So we can see that the tropical paradise is not some pristine forest untouched by humans but is the results of sensitive transitory use over millennia.

There are supposed to be over a thousand uses of that remarkable plant the bamboo. There are more than 40 different genera of bamboo and hundreds of species. Many grow to great heights. Up to 37m has been recorded even though bamboo is an herbaceous species. There is a branched rhizome from which the culms expand like telescopes, from a subterranean base, very rapidly; nearly 1 m per day has been recorded. So important is it that it holds a central place in Chinese and Japanese art and mythology. The first monograph on any plant group was ChubPhu written by Tai Khai-Chih about C.E. 460. Bamboo is said to combine the strength of steel with great lightness. The strength and lightness comes from the hollow tubular structure with fibers around its margin. The hollowness has been utilized as pipes even for piping natural gas. The elasticity of the fibers is a great advantage in regions of hurricanes. The exterior has a thick waxy and silicon impregnated coat that is highly resistant to the weather and to pests and diseases.

The fruits of the forest

Aboriginal Australians exploited a wide range of bush tucker plants, depending upon their homeland. However the types of plants exploited is strongly culturally determined so that people displaced into new areas do not necessarily exploit all the potential plants available. Perhaps the most interesting plants in this respect are those species collected from the wild but also transferred to garden plots like. Horticulture usually includes the collecting nuts and fruits from nature as well as the planting of root plants such as yams and taro. The Kwara'ae collect nuts from *Canarium indicum, Barringtonia, Terminalia, Finschia*, soft fruit from *Eugenia, Mangifera, Spondias, Artocarpus, Paratocarpus, Corynocarpus, Bruguiera, Terminalia, Hornstedtia Rubus, Passiflora* and *Medinella*. For the Kwara'ae the most important fruits are from the Breadfruit *Artocarpus* and *Areca* the betel palm.

Areca catechu, provides betel nuts that are scraped and chewed along with betel leaves (*Piper chavica, P. betel*), and also sometimes nutmeg, cardamom, and lime, as a panacea and stimulant. The lime helps salivation and together the lime and saliva encourage the release of the stimulant arecoline that increases respiration. The leaf stains the mouth deep red. Chewed betel is the stimulant of choice for millions of people from India through to the Pacific Islands.

Artocarpus altilis (Moraceae) Breadfruit, is a fruit treated as a vegetable that has a strong regional and historical importance in Polynesia, even though it is not important in world trade. It had the great advantage of preservability. The starchy fruit was stored in large air-tight pits. A sour semi-fermented pudding produced from the dough called Ma was a staple of Polynesia. Of course it was an attempt by the British to transport such an efficacious plant to the West Indies to feed slaves that spurred the mutiny on the Bounty.

Plantains and banana are the most important tropical fruit by far. *Musa* has about 40 species but the majority of cultivated bananas and plantains come from mainly from just two species and hybrids between them, *M. acuminata* and *M. balbisiana*. Modern cultivated varieties are triploid, they have three sets of chromosomes, rather than the normal two (diploid). Like the staple tropical root crops, they are propagated asexually, and thereby they were released from the necessity to produce seeds; edible mutant varieties with ability to form fleshy fruits without seeds were first selected in diploid *M. acuminata* and then edibility was transferred into a wider range of variants by hybridization as *M. acuminata* was carried out from its center of origin. Different kinds of plantains and bananas have a varying proportion of the parental species. Those with three sets of chromosomes from *M. acuminata* (AAA) are the sweetest and are bananas harvested when they are ripe. Hybrids with one or more sets of chromosomes from *M. balbisiana* (AAB or ABB) are plantains, providing starchy fruit that are generally harvested before they are ripe, and eaten cooked. Edible bananas and plantains originated in Southeast Asia and spread west to East Africa and east to the Pacific Islands at an early stage probably reaching Africa about 1500 B.C.E.

Shifting gardens of the tropics

Tropical horticultural systems have been characterized as slash and burn or swidden systems. Typically small areas of forest are cleared by cutting and burning, but favored wild plants, like palms are maintained. The soil is opened with a hoe or dibble stick to plant the main staple root crops and plantains, as well as many others. The garden is tended, usually not fertilized or irrigated, but weeded and harvested continuously over a long period. The garden looks chaotic because several species are cultivated together, intercropping, but this mixture sustains soil-fertility, prevents erosion, provides a more efficient use of light and moisture and limits pests and disease. The period of cultivation is limited and eventually the plot is abandoned to fallow for several years and another one opened up.

There is a great deal of variation between different cultures in the frequency and extent of the plots and in the periods of cultivation and fallow. Some tropical soils are nutrient poor and on these a relatively short period of cropping of 1-3 years followed by 20 years of fallow is required. This is the case with the Machiguenga of the Peruvian Amazon. They start to abandon their plots after 2-3 years when the decline in soil nutrients gives a poorer return for their labor, at the same time as weeding becomes a more significant labor. However the abandoned garden may still be an important source of food, especially for fruits. Yamomamo gardens have an extended life with plantains harvested for up to five years and peach palms for even longer even after other crops have ceased to be cultivated.

A significant advantage of intercropped gardens is that they provide a variety of products over an extended period, limiting the need for storage of crops or necessitating trade. Labor is spread evenly over the year and so a family can relatively easily maintain its own plot. In all ways family-group is self-sufficient. In the tropical climate fresh roots are available throughout the year. This extended availability is provided by the long crop duration of the major root crops. Cassava is a perennial and may be kept in the ground for 6-24 months. Once harvested it is very perishable. Yams have more storability but also perish, but they have a crop duration of 8 months but may be kept in the ground for up to 18 months. Sweet potatoes have the shortest crop period, of only about 3 months, but availability can be maintained by staggered or sequential planting. Plantains have the lowest storability of all and their availability may be episodic and difficult to predict, with periods of plantain famine alternating with gluts, as different plantings become mature. In the Yanomamo the gluts are occasions for feasting and village get-togethers.

It was probably in the wet tropics that humans first entered into that close relationship with plants, the green betrothal that defines the human species. We are the cultivators, the gardening species. But it was the lack of the necessity for storing foods, their constant availability, that in a way limited the development of human societies in the wet tropics. The next great step in the relationship between plants and humans was not in the wet tropics but in areas where there was only a seasonal availability of the staple plant foods, and so it was that the storability of a food crop became vital. It was not potentially storable tropical crops like yams or breadfruit that were the actors in this drama, but nuts and some weedy grasses, the cereal crops of a seasonal environment. It was not in the lush Garden of Eden that the next major step towards human civilization took place but in marginal habitats where there is only a seasonal availability of food. To see how this might have happened we must take a step back from the horticulturalists, to an earlier stage of human cultural development, to the hunter gatherers.

Hunter gatherers

The !Kung San bushmen of the Kalahari are hunter gatherers who have survived into the 21st century only because they live in an area so marginal that agricultural humans do not want to occupy it. Nevertheless they survive by conserving their resources. They gather only a third of the potentially edible plants available. They are undomesticated humans, living in small family groups, living at low population density, living in temporary camps and traveling over a wide range, following seasonably available food. There are about a hundred edible plants, 40% providing bulbs and roots and 30% fruits but more than 50% of their food is the Mongongo nut from *Ricinodendron rautanenii*.

The !Kung San survive the seasonal availability of food by exploiting different plants at different times but a key to their survival is their reliance on the nut, the least seasonal of their food stuffs. The nut is the fruit of a large deciduous tree. It falls mainly in April and May and matures on the ground. There may be nuts on the ground for up to eight months and it may be for this reason that the !Kung do not store food; there is no need. Nevertheless the !Kung have a low reproductive rate, maintained at a level that is sustainable not just through the normal seasonal variation in the availability of food, but through occasional periods of food shortage because of an extended drought.

The Mongongo fruit is made into a sweet porridge by boiling and removing the inedible skin. The flesh can also be dried for later use. Inside the fleshy layer is a hard stone, 70% of the fruit by weight that is cracked open after roasting to release the tasty kernel about the size of a hazel nut but tasting after roasting, and tasting of cashews or almonds.

The range of plants exploited by the !Kung San is very wide and includes for example the following: Baobab (immature pods, seeds and pulp together) (*Adansonia digitata*); !Hani Vegetable Ivory palm (the flesh surrounding the extremely hard nut) (*Phytelephas macrocarpa*); Kokwaro or Marula (Maroola) (fleshy fruit and "nut" (seed) (*Sclerocarya birrea* subsp. *caffra*); dwarf mobola, mbura or sand apple (fleshy fruit) (*Parinari capensis*); Maua plum, mogorogorwane or Wild orange (fleshy fruit) (*Strychnos cocculoides*); Sour plum (fleshy fruit) (*Ximenia caffra*); Morethlwa and Mokomphata. Raisin bush (berries) (*Grewia* species); Bush mango (fleshy fruit) (*Cordyla africana*); /Tan Root (roasted root); !Xwa water root; Gemsbok cucumber (*Acanthosicyos naudinianus*), Sha root; baboon root (corms) (*Babiana plicata*), Tsin bean (*Bauhinia esculentea*); and Tsama melon (*Citrullus lanatus*).

Even in what seems to us as outsiders an impossibly harsh environment, the desert or semi-desert of the Kalahari, in the most fruitful season they only spend about three hours each day getting food. Apart from ostrich egg shells to carry water the !Kung do not have a means of storage so they do not take more than they can carry. It is dangerous to idealize the culture but it seems lacking the need for permanent storage the !Kung do not hoard material wealth and this removes a major source of conflict. A similar lack of materialism and relaxed life-style has bene described for other hunter-gatherer societies.

The importance of nuts

"A man is the whole encyclopedia of facts. The creation of a thousand forests is in one acorn, and Egypt, Greece, Rome, Gaul, Britain, America, lie folded in the first man." Ralph Waldo Emerson

In temperate and seasonal conditions the relationship between plants and humans is different from that in the tropics. In winter or the season of drought Nature's bounty is withdrawn. If the human population does not migrate to follow the game as they pursue new pastures it becomes essential to store food. Staple foods came from a variety of plants but in many cultures it was a hard seed or nut that was to become a major staple. A nut is a kind of plant food that is self-designed for storage, it carries its own shell to protect the soft nutritious kernel.

Native North Americans, before the European conquest, ate many kinds of nuts including chestnuts, chinquapins, pecans, hickory nuts and walnuts. The staple of the Shoshone people of the Great Basin in central USA was the pine "nut", actually a seed mainly from the piñon pine, *Pinus edulis*. Cones were left to dry and then crushed and shaken in the air to free the seed from the chaff and broken cone. The climate of the Great Basin is one of extreme seasonal extremes, very hot dry summers and very cold wet winters. Late winter and early spring was the season of famine when the many other plants they foraged for berries, grass seeds, roots were unavailable. Pine-nuts harvested in large quantity and stored against this eventuality were the staple food at this time. The population density of the Shoshone was very low in part because the pine-nut harvest was very uncertain, with years of glut and years of shortage. This is a common phenomenon in fruit and nut producers. The occasional years of plenty, called mast years, may be an adaptation of the plants to ensure survival of a few seed at least every few years; there are just too many seeds produced for them all to be eaten. The storage of pine-nuts represents a significant advance for human culture, but nevertheless the Shoshone lived in highly dispersed small family groups for the most time garnering the scarce foods that were available to them. Only in times of glut, and for jack-rabbit and antelope hunts did they come together in larger groups.

The most important nut utilized throughout the northern temperate region were acorns from a variety of different oak species (*Quercus*). In places as wide apart as California, Eastern North America, the Balkans and Japan acorns were the stored staple food. In Japan huge pits were dug to store the acorn crop by the "Mesolithic" Jomon culture. This culture recognized as one of the most sophisticated hunter-gatherer societies that has ever existed. The culture existed from 11,000 B.C.E. to 500 B.C.E. and exploited and managed a wide range of wild foods. It even cultivated in a limited way some species such as gourds and millet, perhaps even rice, though it was the immigration of rice farmers from Mainland Asia that finally brought the culture to an end. However for many millennia acorns were a staple food.

Similarly in eastern North America there was established a culture based on the collection of nuts and the cultivation of local weedy species. It too was converted to more concentrated agriculture by the arrival of maize cultivation from the south. On the Pacific Coast of North America this transition appears not to have taken place, and instead a rich hunter-gatherer culture was maintained up to the time of European contact. For example the Chumash people of California used a wide range of vegetable plants to add to their fish and meat but stored pine seeds and acorns were the staple for the winter period.

There are many species of oak but only a small proportion of these have been used for food. Some species or varieties produce sweet acorns low in the bitter tannins, a fact that Theophrastus, the disciple of Aristotle was well aware. The Holm oak *Quercus ilex*, widespread in the Mediterranean region, is one species with sweet acorns. In some areas sweetness is unfortunately correlated to having a lower fat content and hence less nutritional value. Many species have bitter acorns that require extensive treatment, grinding, leaching and boiling, to remove the highly astringent tannins. The tannins are polyphenols, compounds with multiple hydrocarbon rings that in solution tend to precipitate proteins and hence have an inhibitory effect on enzymes. This is their value to the plant, protecting the plant from herbivores, but unfortunately this also means they inhibit the digestion of proteins and other nutrients. Leaching out of the acorns was carried out by repeatedly pouring water over the meal made of ground acorn kernels. Among North American Indians, acorns were a particular staple of inhabitants of California. They made bread from leached acorn meal made into dough with meat soup or water. Dough was wrapped in fern leaves and placed in the ashes of a fire to cook. Acorns had many disadvantages as a staple, not least the tendency of oak trees to mast, alternating rather haphazardly seasons in which there is massive production of acorns with those in which few acorns are produced. Hence over time they became reserved as food for animals, often by pannage, the herding of pigs in woodlands, or as human famine food.

There were other kinds of nuts, chestnuts and hazel nuts. It is probably no accident that hazel *Corylus avellana* rapidly expanded in Britain about 9,000 years ago, for a while its pollen vastly exceeded that of any other tree species. Perhaps Mesolithic hunter-gatherers traveling with bags of cob nuts to eat and also perhaps sow spread it. Introduced more recently from SE Europe but already recommended by John Parkinson in the 16th century over the native cob nut there is also the filbert, *C. maxima*; it has a longer nut and with a thinner shell. Also grown in SE Europe there is the Turkish cobnut *C. colurna*. In America there is yet another kind of filbert *C. americana*. The prized nature of hazel is emphasized by its importance in magic. Saint Patrick banished snakes from Ireland with a hazel wand.

The chestnut (*Castanea sativa*) was called Zeus's acorn perhaps because of its sweetness; the sativa of its scientific name refers to its edibility. The chestnut is not really a nut but a seed. *C. sativa* is a European species but there is also an American *C. dentata* and *C. pumila* (Chinquapin), Chinese *C. mollissima*, Japanese *C. crenata* and several other species of chestnut. The Japanese chestnut also has a long history of cultivation dating back to the Jomon period. In Europe *C. sativa* had a rapid expansion after the glacial period where it survived in sheltered woodland pockets in the hills in Italy. Its rapid spread about 5,000 years ago coincided with its utilization by Neolithic peoples colonizing Western Europe. From about 3,000 years ago it was cultivated in the Mediterranean region. It arrived in Britain with the Romans. How many centuries have chestnuts roasted over charcoal from a roadside stall been a part of a traditional winter scene? They are served in the same way in Hong Kong as well as an ingredient in a variety of Cantonese dishes.

Two more related genera that provide tasty kernels are the hickories (*Carya*) and walnuts (*Juglans*). Both are found in East Asia and North America and walnuts are also found in Europe and the Near East. Among the hickories *Carya illinoiensis* provides pecans and *C. glabra* pignuts. Walnuts are *J. regia* but several other species are edible including the American *J. cinerea* (butternut) *J. nigra, J. neotropica* and the Japanese *J. sieboldiana*. Walnuts with their brain-like kernel have been praised as brain food. They were a fertility symbol in Roman times.

Water chestnuts

A plant that seems to have played the same role as the Mongongo nut for Mesolithic, hunter-fisher gatherers in some parts of Europe was the Water Chestnut or Water-caltrop (*Trapa natans*). In dwellings dating from 3,900-3000 BC in Latvia deep layers of shells have been found around hearths. It must be well cooked before it is eaten. Its use has been dated back even further to the Early Middle Pleistocene before the emergence of modern humans at an archeological site in the Dead Sea Valley. *Trapa* was just one of many different species harvested by Dead Sea Valley residents for their hard nut or seed and cracked open on stone anvils. The others included another aquatic *Euryale ferox* (prickly water lily), *Amygdalus communis* (wild almond), two species of *Quercus* (acorns) and two species of *Pistacia* (pistachio nut).

Trapa natans is an annual weedy aquatic plant that grows widely in Africa, Europe and Asia in tropical and warm temperate climates. It needs warm water conditions (20° C or more) when it is flowering and produces horny nuts with sharp horns that sink to the lakebed. Either the species is polymorphic or there are several closely related species differing in, amongst other things, the number of horns. Its name caltrop refers to its spininess – literally its name is water-thistle. *Trapa natans sensu* stricto (Jesuit's nut) has four while *T. bicornuta* (Ling nut, Ling gok, Wu ling) from China and *T.*

bispinosa (Singhara nut) from India have two. In China there is another species of "water chestnut" from a completely different family: called *Eleocharis dulcis* or *E. tuberosa*, pi chi or pi tsi is a kind of sedge with a basal corm. It is cultivated in paddy fields. Peeled it can be eaten raw, but it is normally canned, after harvesting it quickly goes off, and as the variety hon matai (ma taai) can be used as a primary ingredient in chop suey, though the Cantonese prefer to use them in sweets, tonic soups and with minced meat dishes. Another variety sui matai is cultivated for starch (water chestnut flour).

At the Latvian archeological site *Trapa* nuts were cracked open with special wooden mallets and then roasted, but unlike some other nut kernels like almonds and acorns that contain toxins destroyed by roasting. The kernel of the nut is rich in protein (20%) and starch (50%). In China it was one of the most important "grains"; in Cantonese cuisine ling nuts are a special delicacy eaten to celebrate the Mid-Autumn Moon Festival. The great value of these hard "nuts" is their storability. In Kashmir they are eaten for about 5 months a year.

Aquatic foods were the staples of a culture that developed across Africa at a latitude that is now the dry southern Sahara or the semi-arid Sahel belt in the period 9000-7000 B.C.E. when conditions were considerably wetter, with for a time many lakes and rivers, including the mega-Chad lake that was more extensive than Lake Superior is today. From about 4000 years ago it gradually retreated until it finally disappeared.

The Arcadian landscape

The range of nuts and seeds is extensive: pine nuts, macadamia nuts, brazil nuts are just a few more. Most have been taken from the wild for millennia. A few have been domesticated but in many cases this has happened only relatively recently. The many years it takes for a tree to grow before it starts producing makes this difficult. Nuts are a wild food, food for free but also disparaged, vegetarian nut roasts a source of comedy. Classical authors have a contradictory attitude to the consumption of acorns. Acorns are seen as one of the principal foods of the mythical Golden Age, Arcadia, the pastoral landscape peopled by shepherd poets when food could be obtained without hard work. The green man or foliate head, surrounded by oak leaves, seen carved in English Churches signifies reverence of the woodlands. This ancient pagan device probably comes from a common source as the Jack of the Green has its root in Celtic folklore.

However in the Classical world the contemporary consumption of acorns in people living on the periphery of the classical world is used as a sign of their ignorance and lack of culture. In history the fate of the acorn was to become the food of pigs, swine. Pannage was the legal right to use the woods in this way. The fate of the prodigal son in the bible, to become a swineherd, is redolent with meaning. Acorns were still utilized as human food but normally only in times of hardship and famine. Juvenal, the Roman poets, in his Satires (XIV, 181) recognized the lesser importance of nuts to cereals *"Acorns were good till bread was found"*. In this satire Juvenal describes how parents teach their children the vice of greed by their example. It is ironic that he uses this example because cereal cultivation and storage is intimately linked with the fall of man, their expulsion from the Garden of Eden, with the human vices and competition that was to result in civilization. The fruit of the tree of knowledge was not Eve's apple but an ear of corn.

It is a mistake to regard the hunter-gatherer cultures as somehow primitive and impoverished. Neither should the hunter-gatherer life-style always be equated with an unsettled or migratory life-style. The hunter-gatherers of the Pacific Coast of North America were an affluent settled society. What makes them different from many other hunter-gather societies was the abundance, sometimes the super-abundance of some of their staple foods including several species of fish, and the development of a technology to store their surpluses. For example there are seasonal runs of candlefish that can be rendered by heating to produce storable oil. The oil was added to other foods, including those from plants, to preserve them. Berries were collected and dried and seaweed made into cakes, each preserved by covering with oil in large watertight boxes made of cedar planks. Along with smoked fish and other animals these provide a large capacity for accumulated wealth, the fuel for cultural sophistication, as well as a cause of jealousy and warfare.

Many hunter-gatherers lived in contact with agriculturalists for millennia but chose not to adopt their life-style. This was the aboriginal people in northern Australia who traded with Torres Island agriculturalists. However other primarily hunter-gatherers groups cultivated some crops and there was a gradual transition to a wholly agricultural lifestyle rather than an agricultural revolution. This was the case in the Americas where cultivation had gradually emerged in the four distinct areas south-central Mexico, north-eastern Mexico, south-western USA and the coastal plain of Peru by about 8,000 years ago, but it was not until 3,500 years ago that settled agricultural communities were established.

In the three northern sites cultivation was based upon cultivation of the amaranth *Amaranthus*. The seeds of wild amaranth are little different from those of cultivated varieties, though these do have larger more productive inflorescences, so that domestication required relatively minor changes compared to say wheat or maize. Similarly peanut was probably the first plant domesticated in South America and it is possible to imagine an easy transition from collecting peanuts to the cultivation of domesticates. Remains from northern Peru have been dated to 6000 B.C.E. but it is likely that the peanut was a staple of the Chilca farmers whose culture dates as far back as 8000 B.C.E. Other plants that was domesticated relatively early were the sunflower in south-western USA and pumpkins in north-eastern Mexico. Very gradually over the millennia the range of domesticated plants was extended to include in different places plants such as Chile peppers, maize cassava (manioc) and potatoes.

However in today's world hunter-gatherers usually live at the margins, where they have been allowed to survive. Almost everywhere they have been replaced by farmers and herdsmen. In Africa the spread of modern farming, and metal working technology, occurred relatively late from about 300 BC with the spread of Bantu-speaking people from central West Africa. Of the two main hunter-gatherer cultures in Africa the Khoikhoi adopted the herding of sheep and cattle but the San kept their traditional lifestyle but as a result were increasingly marginalized.

The rapidity of expansion of the Bantu speakers was remarkable and some have argued was due as much to their cultivation of new crops imported from Asia as their possession of metal tools. The cultivation of plants permitted human culture to develop in entirely new ways, no wonder we talk of the flowering of civilization, though it was many millennia after the first cultivation of plants before this happened. This had to await the cultivation of cereal grasses but perhaps it started with the seemingly small act of people in temperate regions utilizing acorns as a food source. In a fanciful way not envisioned by Ralph Waldo Emerson in the quotation at the beginning of this section, it was in the use of acorns that civilization was sown.

Over millions of years plants and animals had co-evolved, changing each other, and changing the planet. Gardening was a new contract between plants and humans. The human part of the contract involved laboring for the plants; plants provided improved products. In the process some plants were changed, domesticated so that now they relied on humans for their propagation. Some lost the ability to produce seeds. Others lost the toxic compounds that protected them from pests and diseases. In return they were planted and weeded and spread throughout the world in a myriad new varieties.

In the wet tropics starchy root crops and plantains became the staple foods, and palms and bamboos provided for many other uses. A highly developed and sustainable pattern of horticulture was established, a gardening of nature. In seasonal environments hunter-gatherers relied on wild plant products like nuts that were produced in large quantity and were storable over the harsh season. It was when plant cultivation and the storing of plant products came together that the major leap in human culture that we called civilization took place.

2 Bountiful Ceres

"When weary reapers quit the sultry field,
And crown'd with corn their thanks to Ceres yield" Alexander Pope

It wasn't the fruits and roots of the tropical Eden that fed the rise of human civilizations, to the Pyramids of Egypt or Teotihuacan, or the marble columns of the Parthenon, or the Great Wall of China. Neither was it nuts or acorns of Arcadian groves. It was some weedy grasses that started it all. Just the three main cereal crops of wheat, rice and maize, now account for about 60% of human calorie intake and directly more than half of our protein. It is they, along with barley and rye, that started it all. It was Vere Gordon Childe writing in the 1920s who made this connection; between the cultivation of cereals the origins of civilization as if the cereals brought about a revolution in human society. It is now clear that the transition to cereal culture and civilization was slower and more complex than the simple picture Gordon Childe proposed, but it is still true that it was the cultivation of cereals that transformed human society. They have transformed us as much as we have transformed them.

Humans became domesticated at the same time as they domesticated plants. They became sedentary and lived in permanent houses. *Domus* is the Latin for house or home. The houses were grouped in villages and towns. The word civilization comes from *civis* the Latin for townsman. The first humans to cultivate grains ended up living in towns and cities. The cultivation of cereals, and the potential surplus food it provided, enabled populations to grow. They could limit periods of food shortage by storing food, and local deficiencies could be eased by sharing or trading food. Surplus food and larger populations allowed some members of the population to specialise in craft or culture.

The origins of agriculture

The timing of the origins of agriculture seems also mysterious. Agriculture did not first appear until the period between 10-15 thousand years ago in the Natufian culture of the Near East, whereas modern humans had originated about 130 thousand years ago in Africa. Various explanations for this delay have been proposed. The last glacial period came to an end about 15 thousand years ago. The late glacial period and earl post-glacial was a period of uncertain and fluctuating climate, with periods of drought in the Near East. It was here that Palaeolithic people were experimenting with a wide array of food plants, perhaps as famine food, as game became scarce. It was here that the advantages of cereal cultivation for spreading the risk of famine became most obvious. The story, recounted in Genesis, of Joseph's advice to the Pharaoh, to store the surplus of seven years of plenty in order to survive seven years of famine, provides an Old World example of the significance of centralized storage. The Neolithic revolution marked a shift from a ratio of plant foods to meat of about 70:30% to 90:10%.

A climate that is relatively unstable with many minor oscillations in temperature and rainfall is also the kind that favours weedy annual grasses like those that were selected as the first cultivated cereals: wildtype "einkorn" *Triticum boeticum*, wild emmer wheat *T. dicoccioides*, wild barley *Hordeum spontaneum* and Goat face grass *Aegilops squarrosa* and rye *Secale cereale*. Weedy grasses tend to have large grains that allow them to grow rapidly after they germinate to establish themselves and flower more quickly before more slowly growing plants compete with them.

So for humans cereals had the advantage of large grains that a few weeks hard work gathering wild grains, even if collection was by hand-stripping plants, could provide more than enough for a year's supply. Wheat had the greatest productivity of all. Later there was the happy accident of mutations that made the grains more readily harvestable and utilisable that picked out wheat and barley as the first domesticated cereals and led to the gradual abandonment of small-grained grasses: a mutated ear that did not break into bits when it was collected and a mutated spikelet from which grain separated from the chaff easily. And at the same type plants heavily laden with grain were harvested preferentially so that soon cultivated einkorn yielded half as much grain again as their wild ancestors, and cultivated emmer wheat was even more productive.

The history of human civilization in at least three distinct parts of the world is so intimately connected with the cultivation of particular cereals that it is interesting to examine what other characteristics these areas, the cradles of wheat, rice and maize cultivation, shared. These cradles of domesticated plants are regions with well-marked wet and dry seasons. Plants in these areas have to store nutrients either in dormant roots and tubers or in seeds for survival through the dry season. This was the source exploited by Neolithic people. In the post-glacial these regions became more susceptible to drought and human populations became more concentrated beside water sources and oases the margins of which were fringed by the wild ancestors of domesticated cereals. These concentrations of population inevitably resulted in a depletion of wild game in the local area forcing the people to rely more and more on the plants around them, to cultivate the plants as well as harvest them – agriculture.

Looking over the scraped earth of an archaeological dig, the scattering of tiny finds it takes the eye of faith to piece together the family life of a Neolithic farmstead. But so often there is one thing that is readily comprehensible, often the most obvious part of the dig. It is a pit or collection of pits, and the fragments of pottery that are used to date the site. Both signify perhaps the most important aspect of this culture, one that shifted it onto a new level, pottery vessels and storage pits signify the ability to store a food surplus. No wonder these grains became deeply symbolic. Food became synonymous with grain; food in Hebrew is *lehem* and in Greek *sitos* both literally meaning grain or bread. Homer in the Odyssey calls wheat-flour and barley-meal the "marrow of men". The Biblical story of Joseph telling the Pharaoh that there would be seven years of plenty and seven years of famine and to store the surplus in order to survive is an echo of the Neolithic reality. Very much later Jesus was to say "*Do not work for food that perishes but for the food that endures for eternal life*" (John 6:27) echoing the practice of Neolithic farmers.

But the accounting of the costs and benefits of cereal cultivation are not all in the black. Paradoxically the first humans to adopt a sedentary cereal cultivating lifestyle were smaller, about 10 cm shorter, unhealthier, with higher rates of tooth decay and anaemia, and they had to work longer to gain their food than their hunter-gatherer counterparts. So why was the health of the Neolithic agriculturalists relatively poorer than that of their hunter-gatherer ancestors? The short answer to this is that there were many more of them. Paradoxically there was an increase in population even though living together upon the same plot exposed them to disease and parasites to a greater extent. They were more prone to malaria, measles, mumps, chicken pox and smallpox, the diseases associated with contaminated water like cholera and typhoid, and parasites like head and body lice.

The adoption of agriculture required a settled way of life and larger groups, which in turn permitted a reduced spacing between births that more than counteracted the greater mortality. There was less exposure to the risk of famine and child-bearing and women with young children could be more effectively supported in a sedentary lifestyle than a nomadic one. Paradoxically, as we see even today, it is the poorest societies with the greatest infant mortalities that have the greatest rate of population growth; the uncertainties of children surviving to adulthood encourages much closer spacing of pregnancies. Children are of course also much more of an economic asset to sedentary agriculturalists because they can much more effectively contribute to sowing, weeding and harvesting than they can to hunting. Even today in predominantly agricultural societies like India the belief that each child is "one mouth but two hands" has countered attempts to control population growth. Neolithic agricultural population densities were up to 25 times greater than earlier Mesolithic hunter-gatherer cultures.

A comparison of recent and present-day human societies dramatically emphasises the potential for sedentary agriculturalists to have much higher population densities than nomadic foraging hunter-gatherers, even allowing for the considerable variation because of the different potential productivities of their different environments. Hunter gatherer forages such as the !Kung of the Kalihari or historically the Shoshone of the Great Basin of the USA have population densities of about 0.02-0.04 per km^2. Nomadic pastoralists range from very low population densities up to about 3 per km^2 in examples like the Turkana of Kenya or the Kirghiz of Afghanistan. Slash and burn horticulturalists range in population densities relative to the extent of the area utilised and how much the diet is supplemented by wild foods. Perhaps surprisingly the Machiguenga of Amazonian Peru and the Yanomamo of the Venezuelan highlands have population densities of only about 0.3

per km^2; the tropical paradise is not hugely productive. But where there is intensive horticultural practised in the New Guinea highlands and the Pacific Island population densities are typically between 15-50 per km^2. Population densities in agricultural systems based on cereal cultivation exceed these and have risen over the past millennia so that in Japan by the 1700s they were 100 per km^2, and in present day subsistence systems in China and parts of Java they are as much as 200 or 400 per km^2, 10,000 times greater than that sustained by the hunter-gatherer lifestyles of the !Kung or the Shoshone. In the Near East the origins of Agriculture have rightly been called a revolution.

Population pressure encouraged the diffusion of agriculture out from its area of origin. From this distance it looks as if it happened extremely rapidly, like a wave, though it took nearly five thousand years for it to spread from the Near East to the margins of Europe. However in other places the agricultural revolution was even more gradual and more diffuse, an agricultural evolution. In the Americas agriculture originated, probably in isolation, in at least four different places, in coastal Peru, central Mexico, northeastern Mexico and southwestern USA, and there was a long period between the initial cultivation of some plants and the establishment of villages. Interestingly in the last three places agriculture was initially based on cultivation of amaranth (*Amaranthus*) and initially peanuts in Peru (*Arachis hypogea*), supplemented in different places by sunflowers and pumpkins, and it was only later that the American cereal maize was cultivated, and the triumvirate of lowland American civilization maize, beans and squash emerged. Perhaps the delay is related to the extent of transformation of maize, described later, required to make it a useful crop, though once properly domesticated it had huge advantages.

Agriculture has had several perhaps many separate origins at different times in widely separated places around the world but this association between the origins of agriculture and profound climate change is a common. For example the emergence of agriculture in the Andes many of thousands of years later about 1500 B.C.E. is also associated with a period of climate change. The places where climate change was experienced most extremely were those marginal areas where there was a sharp transition between ecological zones, either lowland to upland or wet to dry, where minor changes in climate would radically alter the environment and change its potential for human sustenance. It was in these areas that cereal grasses, or analogue staple crops like potatoes, cultivation, with their potential for storage in an unpredictable environment, that had the greatest advantage[3].

Food surpluses had a profound cultural effect by creating inequalities between individuals, family groups, local groups, tribes and states. Inevitably some people occupied the better land for cultivation, or managed their land better, or worked harder and so had a greater surplus, greater wealth. They could lord it over their less efficient or less lucky neighbors. Food could be granted by the wealthy to the poor to encourage their subservience and loyalty. And grain surpluses could be traded for other material goods that could also be stored as a sign of wealth. Food was power. And now there was a greater motivation for conflict, to steal the material wealth and food of neighbors and surpluses also permitted classes of people, a military class to become skilled in warfare. It was the cultures that developed extensive systems of granaries or storehouses became strong and expanded. In the Americas both the Aztecs and Incas located storehouses throughout their empires in order to support their campaigning armies. They took extensive tribute from subjugated people to stock their supplies.

The cultivation and storage of grain also marked another and more subtle psychological shift. It allowed even encouraged the population to rise but the enlarged population was more exposed to risk. Collapse of the harvest exposed it to famine. Only planning for an uncertain future could mitigate the disaster. Here was the seed of the angst that is at the heart of the human condition. Even while a rich harvest was a joyful event there was the knowledge that it could have failed – *"praise the ripe field, not the green corn"* - the polarity of joy and fear of failure. Here lies the explanation of the religious obsession of early human civilizations with the cycle of the seasons and the astronomical calendar, an obsession with planning, with predicting the future. Here is the seed of the archetypal human cry, the complaint to the gods, why me? It is a dominant theme throughout the Bible. Why have I been made to suffer – externalized to make it comprehendible – externalized by creating the gods. Here is the origin of religion.

We sow the glebe, we reap the corn,
　We build the house where we may rest,
And then, at moments, suddenly,
We look up to the great wide sky,
Inquiring wherefore we were born…
　For earnest or for jest? Elizabeth Barrett Browning

The cultivation of cereals forced human populations down a one-way track towards a certain kind of culture we call civilisation. This has been explained by a rather crazy hypothesis that proposes that the earliest cereal eaters became addicted to their grain, because of the presence of drug-like exorphins within them. This is not say that the first grain cultivator got a drug "high". A better way to think of it is how nice a freshly baked loaf smells, and how difficult it is to give up grains when going on a gluten free or the Atkin's diet. So many human societies today are defined as much by their cuisine as by anything else. There will be more about natural plant "highs" later in this book, but we do not need addiction to explain the success of cereal based agriculture. Once adopted, cereal cultivation allowed greater population densities, densities that could then only be sustained by cereal cultivation; it was a wheel of reliance on the great juggernaut of civilisation. It was the certainty and productivity of cereals, Ceres bounty, that launched human civilization and it was the cultivation of grain that led to the building of the Ziggurats, the Pyramids and the Parthenon.

The human consequences

The Old Testament provides a different perspective on the origins of agriculture. Adam and Eve's expulsion from the Garden of Eden is on one level a record of the transition from the easy life of hunter-gathering to farming. On the third day "Elohim said "Let the earth sprout greenery on the earth, seed-producing plants, fruit-trees producing fruit according to their type" and it wasn't until the sixth day that humans were made and he is given domain of all the riches of Eden to gather. However having eaten of the fruit of the tree of knowledge he is banished. What was the tree of knowledge. Many suggestions have been made. The banana has a suitably phallic shape. Or was it an apple and there is an echo of the disproval of

God because of the alcohol that could be made from it.

So Adam and Eve are banished from the easy life of the hunter-gatherer and adopts the life of a tiller of the soil, a farmer. God tells Adam

"Thorn and thistle will sprout for you when you want to eat the plants of the field: by the sweat of your forehead you will eat bread until you return to the ground (for from it you were taken, dust you are and to dust you will return."

Later it is Cain the farmer's offering to God of the fruit of the ground that is not valued compared to Abel's lamb and God punishes Cain's anger by preventing the ground from yielding its strength. This story certainly reflects the conflict that has always existed between the farmer and the shepherd but is it far-fetched to see a reference to a pre-agricultural paradise and the relative decline of quality of life that the advent of agriculture brought. While God *"planted a garden eastward of Eden"* it was Cain who *"builded"* the first city.

The toiling farmer is an archetype – as God says to Cain

"When thou tillest the ground, it shall not henceforth yield unto thee her strength".

There is a similar conflict in Sumerian literature. The greater gods create a race of slaves to relieve the lesser gods of their labour, dredging the irrigated lands, spreading silt, digging, baking bread. Men were created as slaves to make the gods food and clothes. And like Cain and Abel the god of cattle Lahar and his sister Anshan the goddess of grain were in conflict.

There was a dark side to this revolution. Cereal cultivation and the storage of grain required greater levels of organization and also established inequality. Building and maintenance of the highly productive fields required a large labor force but the "owner" of these lands could pay for this labor. A difference arose between the gentleman farmer and the peasant or slave laborer, a Marxist explanation. There was hierarchy between those with the best land

and the biggest surpluses. They could afford to pay others to work for them and so increase their holdings. They became strong enough to coerce others to work for them. Grain surpluses encouraged trade but also conflict. The stored grain could be stolen. There was a greater tendency to concentrate populations in defensible settlements. Conflict encouraged the development of the state. In war not just grain could be stolen but also slaves could be taken to work the land. So much for the rhyme preached by John Ball in the Peasant's revolt of 1381, "When Adam delve and Eve span, who was then the gentleman," since it was the delving itself that created the hierarchy of gentlemen and peasants.

Cereal cultivation brought toil but it also brought surplus and a social organization. It brought civilization. Agrarian societies are inevitably hierarchical because of the social organization required for grain cultivation; there is the presence of the priest/aristocrat/landowner versus the laborer/serf/slave/lower caste in all important examples of agrarian civilizations from the dawn of agriculture, through the middle ages, through the serf economy of Russia perhaps only ending with the mechanization of agricultural labor. For the cultivation of cereals had to be coordinated, controlled. It is no accident that the greatest civilizations arose first where the greatest central control was required. Technological innovation increased the certainty of Ceres bounty and with this came the first flourishing expansionist civilizations, but with it came systems of control and management.

An urbanized civilization was built upon the cultivation of grain. Grain supplied economies based on the accumulation of surplus food, could also feed non-food producers, craftsmen, soldiers, priests. There was long-distance trade, exposing the civilisation to external influences and having the potential to introduce innovation. A complex division of labour and trade required a record system and the development of a written language. Social differences encouraged the building of large scale public buildings, though in different places these were expressed in different ways – walled cities in Mesopotamia, temples and pyramids in Egypt and Central America. . Economics based on the centralised accumulation of capital through tributes and taxes. Economic support is provided for non-food producers, there is long distance trade and a complex system of division of labour. Advances were made toward record keeping, science, mathematics, the development of a written script, and the construction of public architecture in the form of large buildings.

The most important technology was that associated with the management of water, a hypothesis proposed by Wittfogel. In Mesopotamia and Egypt there was use of irrigation systems supplied by the Tigris, Euphrates and Nile. In Mesopotamia the building of the Hanging Gardens of Babylon is a testament to the availability of a sophisticated irrigation technology. In Egypt the caste of priest kings insured the regular flooding of the Nile by their rituals and made sure the levees and ditches were maintained. In Asia there was also the development of the paddy-field system for cultivating rice. In China and South-East Asia terracing and the development of paddy fields maximized rice production. The availability of metals improved the tools of cultivation. Most of the earliest civilizations arose where irrigation systems were organized.

Some of the most sophisticated water management systems in the ancient world were those of the Indus valley civilization. Water diversion techniques and rain trapping and storage technology extended the availability of monsoon rains. Here in a geographical area as great as western Europe, from about 3,300 B.C.E. an agricultural system based upon the cultivation of wheat, barley, pulses, and sesame, supported at least 1500 settlements. For example the towns had sewerage system that conducted waste to the fields to fertilize them. The decline of the civilization from about 1750 B.C.E. is marked by a failure to maintain the drains and irrigation channels but was perhaps precipitated by one of the major river systems the Ghaggar-Hakra or Saraswati, changing its course and eventually drying.

The Americas provide several different examples of the association between agricultural water management and the development of civilization. The earliest civilization dating from 1800 B.C.E. was that of the Olmec based on the coastal plain south of Vera Cruz. In this zone of tropical forest with a rainfall between 300 cm the alluvial soils are constantly moist and liable to flooding but potentially highly productive. Here the development of techniques to utilize flood prone river-lands and cultivated irrigated river terraces was accompanied by the development of a social hierarchy with a law giving military/priestly nobility over the stratum of peasant farmers and laborers. Like other civilizations relying upon the regular seasonal flooding or rainfall they developed a sophisticated calendar.

The Maya built upon the agricultural technology of the Olmec and were in contact with and traded with the people of central plain of Mexico. On the thin soils over the limestone plateau of the Yucatan peninsula they established raised field-systems that made the best of the seasonal rains. The agricultural system may have supported a population of 3-5 million at its peak, but it was essentially fragile and its failure in a long period of drought so utterly destroyed their ability to feed this huge population that it led to the collapse of the civilization. By 900 C.E. the high Mayan culture had reverted to transient slash and burn agriculture.

The Maya were in regular trading contact with the people of Central Mexico. Here irrigated raised fields were also established to feed the people of Teotihuacan. So the direct descendent of the systems of Teotihuacan, the development of the chinampa system by the Aztecs in the shallow lakes of the central plain can then be seen as a direct descendent of Mayan technology. Whatever, it provides one of the best examples of the causes and consequences of developed irrigated agriculture. With the Aztecs of Tenochtichlan the raised field system became highly developed. The tourist attraction "*the Floating Gardens of Xochimilco*" are a surviving fragment. Strip fields 2-4m wide and 20-40m long surrounded on 2-3 sides by a canal were built up in the shallow lakes by the addition of refuse and vegetable waste. At Tenochtichlan a causeway prevented the freshwater of the lake being polluted by nitrous rich waters during flood. The central Mexican plain has variable rainfall, occasional frosts or very high temperatures, and relatively poorly fertile soils, but by the chinampa system had the potential to maintain year round productivity. The surface of the field was built up by adding vegetation, refuse and silt, and constantly added to thereby maintaining fertility. The water moderated the temperatures. Seedlings were raised in special seed beds and transplanted on. Highly organized, labor-intensive but the system managed to provide more than a half of the food for the burgeoning population of Tenochtiitlan of more than 200,000 people and support large castes of craftsman and artists, merchants, priests and nobles.

There is a similar link between the introduction of maize and the establishment of extensive irrigation in South America and the rise of Peruvian and Andean civilizations. Small villages of hunter-gatherers, fishermen and potatoes and bean gardeners were established along the river valleys of the north coast of Peru by about 2500 B.C.E. but about the 1300 B.C.E. maize was introduced (i.e. about the time of the Olmec civilization in Mexico). Extensive field systems and irrigation works were established. For example, the Chimu Intervalley Canal of 950-1100 CE brought irrigation water from the Chicama River to Moche Valley field systems surrounding the ancient capital at Chan. Irrigation and rainwater management system have been dated back to 1500 B.C.E. Another interesting example comes from the high altitude Andes again involves water management. Cultivation at 4500 m altitude around the city of Tiwanaku in Bolivia dating from 600-1100 CE took place in raised planting platforms drained by surrounding shallow canals. The fields that garnered the suns warmth, and limited water-logging in the perennial wetness of the altiplano.

But all agricultural systems were not dominated by considerations of water management. Dating from about the same period as the Aztecs and a quite different but also intensely hierarchical system was the strip field system of the English Middle Ages that provided the economic basis for the flowering of Early Middle Age culture. There is only one direct record for the cultivation of scattered strips in the Domesday Book of 1086, that for Garsington in Oxfordshire. How much this is because it was so common as to be unremarkable or because it was actually rare is difficult to know. Nevertheless the cultivation of strips within a vast open field had become the normal pattern of cultivation by the thirteenth century. The system was so successful that it provided food as population rose rapidly so that by 1300 C.E. at the height of medieval development it was 5-6 million, concentrated in a great swathe across the Midlands into Lincolnshire and East Anglia. Since by far the majority of the population was working the land, the countryside must have been teeming with people. The area with greatest population, the Midlands and East Anglia, was also the area where there was the greatest development of the open field system. In the Domesday Book arable land was measured by the number of plough teams and already in 1086 these were concentrated in the Midlands and East Anglia where there were more than 3.5 plough teams per square mile. The number of plough teams quadrupled in the next 150 years.

There was not one system, open field cultivation did not conform to one pattern, even within the area where its development was greatest. Generally there were 3 vast open fields, there could be 2 or 4, each divided into rectangular blocks or "furlongs" which were subdivided into individually owned strips. A typical strip had an area of about one acre, about 200 by 20m (1 furlong by 1 chain = 220 yd. by 22 yd.) and could be ploughed in one day. It has been suggested that strips originally arose as square fields about a furlong were divided between the sons. A strip often consisted of a several adjacent plough ridges, selions, about 5m wide. These may have initially been raised with spades to aid drainage, but the method of driving the plough clockwise around the ridge turned earth towards its center. The ridge was about 1m higher than the furrow. The ridge was oriented down the slope to improve drainage.

The strips were tenanted though most work was done communally. There were generally three classes of peasants: free men or franklins, villeins, husbonds or neats who were the most numerous, and farmhands called cottars, cotterells, bordars or undersettle. The holding of a freeman or villein was about 12 hectares composed of scattered strips and proportional shares of meadowland and grazing rights on the common. The open field system was a sophisticated kind of land management which depended upon a balance of fallow, manuring and rotation of leguminous crops which maintained fertility and controlled weeds. Strips were not separated by fencing or hedges except perhaps at the boundary of the field. Laws were laid down for the compensation of those who lost produce through the failure of someone to maintain his part of the common fence. The system was governed with by-laws, for example

"...no one shall gather beans, peas or vetches in the fields who holds land of the lord, except from the land which he himself has sown".

The rotation of crops in close vicinity provided ample opportunity for different kinds of weeds. As well as fallow land there were extensive headlands, where the plough turned round, which were raised from the accumulation of soil pushed ahead of the plough share. There were also gores, awkward angles of land between strips. These areas which were frequently disturbed must have had a diverse ruderal and weed flora as well as stray crop plants. These areas could also be harvested. One by-law allowed

"...that a pauper shall gather beans not inside, but at the head or alongside the selions. And if they do otherwise they are to surrender whatever they have gathered and not be allowed into the fields to gather beans again..."

Cultivation of the same piece of ground had the twin disadvantage of declining soil fertility and increasing weed infestation. The introduction of regular rotation favored some weeds over others. Rotation of crops was practiced, but in different ways in different places. It could occur by the whole field, so that only one kind of crop was growing in the fields at a time. Alternatively the field would have some furlongs under winter wheat or rye and others with spring sown barley, oats, peas, beans or vetch. Each time weeds could re-establish from the soil seed-bank. A seed bank of weeds and grassland species can persist for many years in a soil cultivated continuously. Weed infestation must have been a very serious problem in the open field. Henry II issued an ordinance against Guilde Weed *Chrysanthemum segetum* (Corn Marigold) and legislation was also enacted against it in Scotland. In 1597 Gerard describes Corn cockle with these words

"What hurt it doth among the corne, the spoile of bread, as well as in color, taste, and wholesomeness, is better known than desired"

An alternative name for Cornflower was Hurt-sicle. Gerard describes the way Cornflower
"hindereth and annoyeth the reapers, by dulling and turning the edges of their sicles in reaping corne."

A particularly problematic weed was *Anthemis cotula* (Stinking Chamomile). An alternative name for it is Stinking Mayweed from the Anglo-Saxon *"maegthe"*. It became so common in some areas that it gave its name villages like Mayfield in Sussex and Maytham in Kent. Fitzherbert in his *Boke of Husbandrie* of 1523 describes *"mathes"* or *"doggefenell"* as *"the worst weed that is except terre"*.

"Terre" is probably *Vicia sativa*. Stinking Chamomile gained its bad reputation in part because of its ripe achenes could cause the skin to blister. It flourished in the open field because of its genetic diversity and plasticity in growth. It could produce vigorous new shots from the surviving basal part of the stem after the top was scythed off and it could also grow as a winter annual. One way to control weeds was to allow grazing on the field after harvest. However *Anthemis cotula* was relatively trampling resistant and unpalatable to grazing animals. This method of weed control could give rise to problems on neighboring strips. A by-law of Newton Longville said

"...no one shall cause his beasts to graze in any piece of cultivated land before the crop of at least one acre adjacent is wholly removed".

In addition, at any time, one whole field was left to fallow. Cattle were grazed on the fallow field. A period of fallow helped to control weeds. The folding of grazing animals on the fallow field allowed the trampling of weeds and manuring to be concentrated in particular areas.

"The *great intention of a fallow is to pulverise the land and destroy weeds*" - Pitt 1809.

In the fallow, weeds were grazed back and trampled, not allowed to flower by being ploughed in green in early summer, suffered competition from more vigorously growing plants in longer fallow. At first fallow was allowed to re-grass naturally but later the fallow was sown with seed from hay. Grasses like *Lolium perenne* (Perennial Rye-grass) and *Cynosurus cristatus* (Crested Dog's-tail) provided the most important grazing.

Wheat and Barley, the staff of life

There is archaeological evidence that wild barley was being harvested 10,000 years ago. It became the most abundant and cheapest grain in the Near East. Like wheat there are hulled and naked, and brittle and non-brittle cultivars of barley. Wild types of *Hordeum spontaneum* have two rows of hulled seeds on a brittle ear that shatters into segments when harvested. The earliest domesticated variants *Hordeum vulgare* had a tough ear that stayed together when harvested and by 6000 BC cultivars had naked grains and a six-rowed spikes.

At first it was grown in upland areas which had enough seasonal rainfall to nourish the crop, but soon both in Sumer and Egypt it was taken down into the lowlands, where it was cultivated in irrigated fields. Here on irrigated fields very rapidly growing varieties arose; the Sumerians describe a 30-day variety among others. It was the favoured staple cereal in Mesopotamia and Egypt and remained the most important cereal up to the time of the Romans. In the bible it is barley that Ruth gleans from the fields.

Cultivation spread widely reaching China in the second millennium BC. Barley is not as fastidious as wheat, germinating more readily in drier conditions, growing on the thin limestone soils widespread in the Mediterranean region and maturing more rapidly. Barley also had the advantage over wheat of being slightly salt tolerant so that on irrigated lands in the Lower Mesopotamia that had a tendency to become salinified, it had completely replaced wheat by about 1800 B.C.E.

By late classical times barley was regarded as the food of the poor. Jesus feeds the five thousand on barley bread. This poor reputation may be based on low gluten content so that it produces rather poor unleavened bread. Normally barley groats were roasted first and eaten in a porridge, polenta or gruel. Nowadays polenta is normally made from maize flour. The Greeks roasted the soft barley grain in a special dish called a *phrygetron*. The roasted barley could be eaten or ground to make barley-meal. The meal was used to make barley-cakes, maza or bannocks. They had a widespread reputation in classical times as a source of strength. Spartan boys were fed on them. Later gladiators, were known as *hordacei*, barley-men. However by Roman times barley was regarded definitely as a second best food. Roman troops were punished by being fed on barley rather than wheat. Its major use today, apart from providing animal feed, is for brewing-malts, more of this later, but this may account for its rather paradoxical preference as a gift for the gods.

The earliest archaeological evidence for the domestication of a cereal is of wheat. Charred remains of domesticated einkorn and Emma wheat, and a little later barley, have been identified from the remains of Neolithic villages of about 11,000 years ago in the area of western Turkey, Iraq, Syria and Israel. It was several thousand years before an agricultural economy permitted to rise of the great ancient civilizations of Mesopotamia, Egypt and the Indus Valley, by which time agriculture is embedded in myth and religion.

Unlike barley that could be grown anywhere, the cultivation of wheat in classical times was patchy so that wheat was an important part of trade. Greece obtained wheat from the plains around the Black Sea and also from Egypt. It encouraged the founding of Greek colonies around the Mediterranean and Alexander's conquest of Egypt. After the annexation of Egypt by the Romans large quantities of wheat were shipped to Rome. Wheat was named *Triticum* by the Linnaeus after the Roman name of bread wheat, a name derived from the Latin verb to grind *terere*.

As wheat cultivation spread new cultivars arose through mutation and hybridization between species in the genus *Triticum* and a closely related one called *Aegilops*. Today there are six species of *Triticum* recognized, five of which have cultivated varieties. They differ in their number of sets of chromosomes, two, four or six, or diploid, tetraploid and hexaploid, with either 14, 28 or 42 chromosomes. There are thousands of cultivars. New cultivars were successful because they were adapted to different day-length or were day-neutral, and so could be grown anywhere and at different seasons, or had a different seasonality, and different tolerance of rainfall, temperature and different soils. One important change was an increase in the number of grains in each spikelet of the ear from one (hence "einkorn") to two to several.

One diploid species called *T. urartu* is only found in the wild. *Triticum monococcum* has a wild subspecies widely distributed in the Middle East called *T. monococcum* subsp. *aegilopoides* (*T. boeticum*). Cultivated *T. monococcum* differs from the wild sort by having a stalk of the wheat ear (the rachis) that is tougher and so allows effective harvesting. Instead of hand-stripping plants before the grain has fallen, or picking up individual grains from the soil after they have fallen, a handful of wheat stems bearing their ripe grain could be cut with a sickle. However Einkorn is primitive in one respect; the grain kernel is surrounded by bracts or hulls. Cultivated Einkorn *Triticum monococcum* is closest to the ancestral domesticate, and the one found in the stomach of Otzi "the Neolithic

"Iceman", who emerged from his resting place in an Italian Alpine glacier in 1991. Cultivated einkorn that is closest to wild einkorn is found in the region that lies between the sources of the Tigris and Euphrates in eastern Turkey. A more derived einkorn called *T. sinskajae* by some taxonomists is hull-less so that in threshing the hulls are freed ("free-threshing") and a higher quality grain is produced. The cultivation of Einkorn is very rare today though it has been found in scattered localities from Spain to Turkey. In some places it was until fairly recent decades used to make porridge and in others it is still used as food for pigs.

Einkorn did not give rise directly to other cultivars. Rather these arose after hybridization between *T. urartu* and a species called *Aegilops speltoides*) with subsequently a doubling of the chromosome number in the hybrid to make a fertile hybrid species. This kind of hybridization was encouraged by the overlapping geographical distribution of *Triticum* and *Aegilops* species. Archaeological examination of storage pits has sometimes indicated that mixed cultivation of several varieties or *maslin* was being practiced in some places. *T. timopheevii* is another derived hybrid species of *A. speltoides* and *T. monococcum*. It has a wild subspecies called *T. timopheevii* subsp. *armeniacum*. Domesticated *T. timopheevii* has a very minor importance today and is only cultivated in a few places in Georgia. It also by hybridizing with *T. monococcum* gave rise, after doubling the number of chromosomes to the hexaploid *T. zukhoskyi*, whose cultivation is also limited to Georgia.

The most widely cultivated wheat in the early farming villages of the Near East, was another derived hybrid species called Emmer wheat or *Triticum turgidum* subsp. *dicoccum*. It is non-brittle but has a hulled grain. It gave rise to durum wheat *T. turgidum* subsp. *durum* that is free-threshing and widely cultivated today. Nevertheless hulled emmer continued to be widely cultivated because it was hardier than durum wheat, and in storage more resistant to pests and disease. It retained its cultural significance and in Roman times emmer potage or *puls fitilla* was used as a sacrifice.

Triticum aestivum also has a hybrid origin but between emmer wheat and a wild diploid species called *Aegilops tauschii*. This wild species introduced a tolerance of more continental climates, more extreme climates, and permitted cultivated wheat to spread into central Asia and India by the beginning of the third millennium BC. Today it is *T. aestivum*, the bread wheat that is cultivated worldwide on a vast scale. *T. aestivum* subsp. *aestivum* wheat has become adapted to a wide range of conditions though it performs best outside the tropics,, optimum temperature is 25°C, with minimum and maximum growth temperatures of 3° to 4°C and 30° to 32°C, and 75% grows in areas with between 375 and 875 mm of annual precipitation. Commonly spring and winter wheat are distinguished. Winter wheat is sown in autumn and takes advantage of autumn rains and early spring sunshine; it requires a period of winter coolness (0° to 5°C) before it flowers. Another variant is *T. aestivum* subsp. *spelta* (*T. spelta*) provides spelt flour is said to be more easily digested than bread wheat flour. It is hulled and contains gluten but it is becoming increasingly popular as a health-food crop because although it contains gluten it is said to be more suitable for many people who are intolerant of the gluten and it has a higher protein content.

The Romans were fond of a gruel called *apotherium* made with whole grains of spelt. The recipe according to Apicius the Roman gourmand and cookery writer from the first century C.E. seems to have been a kind of wheat pudding rather like rice pudding.

"Boil spelt with pine nuts and peeled almonds in [boiling] water and washed with white clay so that they appear perfectly white , add raisins, [flavor with] condensed wine or raisin wine, and serve it in a round dish with crushed [nuts, fruit, bread, or cake crumbs] sprinkled over it."

The modern version in Cuccìa a Sicilian pudding eaten on Saint Lucy's Day, the shortest day of the year in the Julian calendar, a day on which the eating of anything made with wheat flour was banned. Cuccìa is said to commemorate the arrival of grain ships ending a famine, but must hark back to pagan rites of rebirth. In Ancient Greece a similar dish made with whole grains, and called panspermia, celebrated Apollo and the coming and going of the summer sun.

The utilisation of cereals was not straight forward. There is an extraordinary diversity of products because wheat varieties differ in their chemistry, and they can be treated in many different ways. Wheat is perhaps the most palatable of all cereals but it had to be freed from the chaff by threshing and winnowing. Hulled varieties still have the outer bracts of the wheat flower, the chaff closely surrounding the grain and result in a higher fibre product. Most modern varieties are free-threshing, meaning that the chaff separates easily and can be removed easily by threshing and winnowing. Then the grain, sometimes called a berry, but more technically a caryopsis, the seed inside the fruit wall called the pericarp, is milled and sieved.

Primitively milling was carried out very laboriously between two stones. A saddle quern was the earliest form: the large lower stone was rectangular with a concave upper surface in which grain was rubbed backwards and forwards by the smaller rounded top stone. In classical times they were replaced by rotating querns had two rotating stones the lower one concave and the upper convex with a central aperture through which the grain was fed. The beehive sort had a hemispherical upper stone with a slot for a wooden handle but milling became increasingly a commercial operation using large cylindrical stones rotated by donkeys or mules. The millstones were made of hard rock like basalt but grooved. Watermills originated in China by about 400 C.E. Windmills may have originated in Persia. By varying the gap between them different qualities of meal could be produced from cracked wheat through coarse grits or groats, and polenta, to finer grades. Coarse meal could be made by pounding too; in Rome groats was made by pounding the grains and Roman bakers were called *pistores*, pounders. Similarly coarse milled barley is called pearl-barley and finer milled barley called barley-meal.

After milling the flour is sieved and a variety of grades of flour produced. Milling frees from the starchy endosperm to a greater or lesser extent, depending upon the variety, the outer layers of the grain or the bran, which includes the fruit wall or pericarp, the seed coat or testa and the aleurone layer which mobilises the starch in the endosperm on germinating,. Sieving gradually reduces the proportion of protein and fibre and increases the percentage of starch, and makes the flour whiter. A major part of the trace elements for human health phosphorus, magnesium, chromium, zinc, and manganese is located in the bran. The content of trace elements also depends upon the conditions of cultivation especially the nature of the soils. Milling does have an important advantage that surviving nutrients such as iron in the flour are more available for human nutrition, because of the reduced concentrations of complex polysaccharide and polyphenolic compounds such as tannin and phytate that are known to limit its ability to form ions. On germination phytate, and other anti-nutritional factors, are broken down, so malted grain has an enhanced nutritional value.

In recent years concerns have been expressed that because the source of wheat flour used in the UK has shifted from Europe rather than the USA and Canada, and does not provide an adequate supply of selenium. Selenium is incorporated into antioxidant enzymes the selenoproteins such as glutathione peroxidase, which help prevent cellular damage by scavenging the damaging free radicals that are natural by-products of oxygen metabolism. In doing so, they limit the kinds of damage that may lead to heart disease and cancer. Selenoproteins also play a role in the immune system. In some parts of China and Russia the daily intake of selenium is less than 10µg compared to 170µg in selenium rich parts of the USA and Canada, and there it is the deficiency is implicated in a high rate of viral diseases and a heart disease called keshan. The importance of an adequate intake of selenium has been emphasized to people with HIV/AIDs. Luckily only a very little selenium is required, indeed it is toxic in high concentrations, and the little required is easily obtained by eating a few nuts like Brazil nuts, or by using selenium rich fertilizers on the crop.

Wholemeal includes the bran, the embryo and the endosperm and wheat meal that has some bran blended with white flour are recommended by health professionals not just because of their higher concentration of trace elements but also because of the fibre and arrange of phytochemicals like lignans that they provide.

The description of wheat flour types is complex because of the range of species and varieties used and the variety of processes involved in the preparation of the flour. The chemistry of the grain is the most important factor determining how the flour can be used. The key ingredient is a protein called gluten, actually a mixture of proteins called prolamines and glutelins called glutenins in wheat. One glutenin called giladin is the major cause of celiac disease in susceptible people, but it is the gluten that allows bread and cakes to rise; it traps the carbon dioxide produced by yeast fermentation. About 10% of the endosperm protein is made up of high molecular weight (HMW) glutenins. These confer the elasticity to dough trapping air in baking. The more regularly the amino acid units repeat and the more often the polypeptide chains cross-link by disulphide bonds the more elastic the dough. Weak flour, roughly equating to cake flour, has a low gluten content (<8%) while strong flour has a high gluten content, and is good for baking leavened bread. *T. durum* wheat contains 8 different genes for glutenin proteins and *T. aestivum* is even more potentially variable with 12 different genes. Although some of these genes might have identical alleles there are at least 16 different alleles found together or separately in different modern European wheat varieties and there was probably even more variation earlier in the evolution of wheat. So there is a great potential for different qualities of flour.

Both emmer and durum wheat flour can be used to make bread. The Egyptians made leavened bread from emmer. Emmer flour is lower in protein and more easily digested than other wheat flours. *Farina* the Latin word for flour comes from *far* the Latin word for emmer. To Italians emmer flour is *farro*. The Roman national dish, was a porridge called *puls*, made from the groats of hulled emmer. They also made dried little dough balls from called *tracta* from finer milled emmer, and used them to thicken soups and sauces. By the time of Imperial Rome bread (*panis*) was being eaten as much as porridge though they did not regularly add a leavening agent. Cato distinguishes *panis* from *panis depisticus* (kneaded bread). Pliny the Elder describes several wine-based means of leavening the bread, but it seems that it was the brewing cultures with their ready supply of brewer's yeast that most regularly leavened their bread. Emmer is still used in soups, pizza crusts, breads and cakes.

Durum is harder to mill hence its name Today durum wheat is widely grown in drier areas and provides grits (groats) for couscous, coarse semolina flour or macaroni flour used to make pasta and some kinds of noodles. The high protein and hard starch content holds the pasta together in boiling water. Pasta may have first come to Italy through Sicily. Noodles were an ancient Middle Eastern dish, called *rishtu* in Persia, vermicelli was *itrya* in Arabic. Pasta and noodles had the enormous advantage of pre-preparation, so that once made and dried they could easily last 2-3 years.

Between 70 and 75% of the wheat grain is starch, made up of amylose and amylopectin. The milling process can alter the behaviour of the starch and hence its cooking qualities. Damaged starch grains interact with other ingredients and may reduce the ability of a dough to expand. Soft wheat flours are less likely to be damaged and therefore favoured for the baking of cakes. One variable qualities of the starch is its ability to swell and its stickiness. High starch swelling makes for soft noodles. There are also flours with high α-amylase contents, that are liable to convert the starch to sugar and so useless for baking though very good for alcoholic fermentation.

Another variation comes from pre-cooking the grain. Bulgur (*burghul, bulger, bulgar, wheat groats*) is made by steaming wheat kernels and then drying and crushing them. It is common in the Near East and Eastern Mediterranean region and dates back at least 4,000 B.C.E. Bulgur has a nutty flavor and has a long storage life, resisting fungal mould and insect attack.

Rice

The history of the cultivation of rice provides another example of the importance of social organization. The first domestication of rice occurred a little later than wheat but probably about 6000 B.C.E. Rice was only one of several types of perennial lowland grass domesticated in the broad region from north-eastern India and the coast of southern China. This is the center of diversity of rice cultivars today. There are two related progenitors in this area, the deep-water perennial *Oryza rufipogon* and *O. nivara* an annual species that grows around waterholes and along ditches. A separate domestication involving different ancestors the perennial *O. longistaminata* and the annual *O. barthii* took place in Tropical West Africa.

Domestication gave rise to involved the familiar shift to larger grains and a non-shattering inflorescence, and here as well was a shift to the annual habit and shorter dormancy. Pottery shards bearing the imprint of both grains and husks of modern rice *Oryza sativa* were discovered at Non Nok Tha in the Korat area of Thailand dating from 4000 B.C.E. Already by this time intensive slash and burn agriculture for rice cultivation had been established in the

Middle and Lower Yangtze valleys.

Early cultivation was of upland or dry-land rice, cultivated on free draining slopes. This remained the predominant pattern of cultivation in SE Asia, its putative center of domestication, for several millennia. The kind of pattern of shifting rice cultivation practiced by the Iban of Sarawak where the rice field is cultivated for only one or two harvests and then a new plot is cleared from the forest.

The spread of rice from its proposed area of domestication in SE Asia northwards into China depended upon the evolution not just of a weaker photoperiod response allowing flowering at a northern latitude, but also to some extent on changes in agricultural practice. Puddling of soils extended soil moisture availability and transplanting of 1-6 week old seedlings in paddies gave rice plants a start over other plants and increased yield. Subsequently wetland cultivation spread by the migration of peoples to other areas in southeastern Asia including into SE Asia. Wetland cultivation is so extensive now that methane-producing bacteria in rice-fields are suspected of putting 115 Mt of methane, equivalent to all the world's natural swamplands, into the atmosphere each year adding to global warming. The introduction of drought resistant and earlier maturing rice from Champa, a Malay kingdom in southeastern Vietnam, doubled the area of rice cultivation in China in the reign of Zhengzong (980-1022 C.E.) by allowing well-watered slopes to be cultivated as well.

Hindus believe that the Lord Vishnu caused the Earth to give birth to rice, and the God Indra taught the people how to raise it. In China a proverb is "the precious things are not pearls and jade but the five grains". Rice is the foremost of these. Complex rituals and superstitions surround the planting of the rice crop. Among the Miao people of south-western China rice must be transplanted in odd numbers.

There is no better example of the importance of cereal cultivation in the development of human civilization than the history of rice cultivation in Japan. In Shinto belief, the Emperor of Japan is the living embodiment of Ninigo-no-mikoto, the god of the ripened rice plant, and the last emperor Hirohito religiously cultivated his own rice patch in the grounds of the imperial palace. Rice cakes, *mochi* and rice-wine, *saki*, are an integral part of many Shinto ceremonies.

Rice cultivation has been associated with the transition to the Neolithic Yayoi culture in the 3rd century BC. The evidence about cultivation is equivocal about the earlier Jomon culture: there is some evidence of millet, and even rice, cultivation but it seems the Jomon people behave mainly in a Mesolithic manner, surviving by hunting and gathering. Never-mind with the introduction of short-grained rice from Korea, to northern Kyushu and the establishment of new patterns of wet cultivation of rice Japanese civilization took off. Short-grained rice is now called "Japonica" rice.

An early name for the homeland was *mizu ho no kuni*, the land of the water stalk plant (rice). The cultivation of rice is deeply engrained in Japanese society: *Gohan* is the word for cooked rice – extended as *asagohan* for breakfast, *hirugohan* for lunch and *bangohan* for dinner. The cultivation of rice transformed the society from one of hundreds of scattered tribal communities to a unified society. Rice cultivation is labor intensive and requires co-operation to maintain irrigation and share water supplies, linking families in a harmonious codependency; *wa*, consensus-seeking is a fundamental aspect of Japanese society even today. However, as with cereal cultivation elsewhere, it led to social differences; the headman had the largest store of rice, wealth was counted in rice measures, *sho*, and for many centuries rice was used to pay taxes. Even the samurai were paid in rice.

Similarly among the Iban of Sarawak shifting cultivation is the cooperative venture of the longhouse and is associated with a rice cult and rituals. Rice seeds possess souls so that the planting of rice is a highly significant act. One of the two most important gods is Sempulang Gana god of the earth and rice cultivation. Sake, or rice wine, is used to purify temples in the Shinto religion and in a wedding rite called Sansankudo the bride and groom drink sake.

It was advances in the breeding of new cultivars of the grains, wheat, maize and rice that brought the green revolution of the 1960s. New cultivars of rice had a shorter stature with a greater allocation of energy to the grain rather than leaf, and higher responsiveness to nitrogen based fertilizers. The new varieties, one particularly successful one was called IR8, had the potential for much greater yield and were day-neutral so they could be grown at any latitude. At the same time new varieties of wheat enabled the enhanced cultivation of wheat in drier areas such as Pakistan and Mexico.

However there were negative consequences. Some of the new varieties were not like because of aspects that had not been identified in the breeding programme. IR8 was disliked for its cooking quality. Some new wheat varieties were red rather than amber in color. Perhaps more importantly there was a loss of genetic diversity as long cultivated landraces were abandoned. The new varieties were actually part of a package that included new methods of cultivation. Although yields were greater so were inputs especially in fertilizer.

A new variety of rice has been at the center of the GM food debate. First in 1977 genetic resistance to rice blight caused by *Xanthosoma oryzae* pv. *oryzae* was identified in a wild rice from Mali called *Oryza longistaminata*. The introduction of this resistance into cultivated rice by traditional genes would have taken many years of crossing and back-crossing. Instead the gene was isolated and cloned in a microbe to multiply it up. It was then introduced into crop rice by coating gold particles with the gene and firing them into a rice cell culture using a helium powered gun. Alternative methods of introducing exotic genes include the use of genetically modified plant viruses like the tobacco mosaic virus as the transporter.

Then in 2000, in the middle of the debate upon the safety and environmental impacts of GM foods the biotech company Zeneca announced the development of a new variety of rice, "Golden Rice". It was a cultivar of rice engineered to produce beta-carotene, giving its eponymous golden color, that allowed people who ate it to produce their own vitamin A. Since the consumption of green vegetables, the main source of vitamin A, has been in decline worldwide, and the World Health Organization has estimated that 250 million people worldwide are deficient in vitamin A, increasing their risk of blindness, suppressing their immune system, and causing other health problems, golden rice seemed to be a tremendous advance. Zeneca promised to offer the seeds to poor farmers for free. However dissenting voices were raised. First that unfeasibly large quantities of golden rice would have to be eaten to get the recommended daily dose of vitamin A but more importantly the suspicion that golden rice was only a Trojan Horse and was being used to gain acceptance of GM foods.

Other grains

Within the regions once covered by seasonal and savannah grasslands a range of other cereals are grown especially sorghum and various kinds of millets. Millets and sorghum differ in the form of the grain. In sorghum and pearl millet the fruit coat, the pericarp, is fused to the seed forming what is called a caryopsis. In other millets it is not, and the grain is called an utricle. Milling is more likely to remove the pericarp in these sorts. Starch is the major storage form of carbohydrate in sorghum and millets, and consists of a mixture of amylopectin, a branched-chain polymer of glucose, and amylose, a straight-chain polymer. The digestion of the starch depends on hydrolysis by pancreatic enzymes but processing of the grain by methods such as steaming, pressure-cooking, flaking, puffing or micronization of the starch increases the digestibility by releasing of starch granules from the protein matrix rendering them more susceptible to enzymatic digestion. Waxy, glutinous or sticky sorts of starch have a high content of amylopectin and are more digestible. Floury sorts are high in amylose. Some varieties of sorghum and millets are 100% amylopectin.

Foxtail millet comes in two sorts, sticky and non-sticky. Sticky types can be used for casseroles and pizza. Non-sticky types apparently tastes like minced chicken and as a flour can be used to make pancakes. Foxtail millet is nowadays regarded as the food of the poor, and is mainly used as animal feed but this was not always the case. Foxtail millet is one of the five sacred grains of China (along with barley, wheat, rice and soybean). Foxtail millet (*Setaria italica*) was being cultivated on the floodplains of the Huang Ho (Yellow) River between 7000 and 6000 B.C.E. somewhat earlier than rice cultivation was established further to the south in the region of the Yangtse. On the floodplains the deep loess (windblown silts) blown from the interior of Asia is cut through by the rivers, but provides a rich deep soil. Here foxtail millet provided not just a relatively fast maturing grain but straw for fodder and fuel. Foxtail millet cultivation spread widely especially to India and Japan where it is still an important crop today. Until the *Meiji Ishin* revolution in 1867 millets were more popular than rice in Japan. The cultivation of foxtail millet spread even as far as Europe where it appears in archaeological deposits of Neolithic Lake dwellers.

In this part of northern China brassicas, soybean and hemp were also cultivated in a mixture of vegetable garden and slash and burn temporary fields. This Yangshao culture exhibited many of the aspects we would now consider as especially Chinese including the cultivation of silkworms. A sophisticated village culture with the production of fine ceramics and leather and basketry was established. Hillsides were terraced and drainage ditches established. Like other major centers of origin of agriculture the area was climatically sensitive and there was a marked decline in the culture in between 4000-3000 B.C.E. in the Guanzhong Basin can be attributed to the climatic aridity at the time. Rice cultivation arrived from the south about 3000 B.C.E.

Sorghum (*Sorghum bicolor*) has also been identified from Neolithic sites in the Huang Ho region of China though it is a normally regarded as having a tropical origin probably in the region of northern Sudan westward, what is now part of the southern Sahara Desert but was 10,000 years ago a much wetter region with strong seasonal rainfall. Here there was a gradual transition from a nomadic pastoralism, with the early domestication of cattle, supplemented by wild grains, to a settled life-style reliant on wells and with the construction of granaries for storage of domesticated sorghum, a transition that had occurred by about 9000 years ago.

In other parts of Africa, along the margins of the Nile and its tributaries and around Lake MegaChad there was a related much denser population reliant on hunting and foraging aquatic sources of food but later the shrinking of aquatic sources of food, with the retreating shore-line of Lake MegaChad and the drying of Nile tributaries, encouraged the Sudanic people to turn more to a mixed pastoral-farming lifestyle in which sorghum cultivation played a more significant part. Thus cereal cultivation and agriculture quite clearly have an independent African origin that may have been just as influential in the rise of Nile based civilizations as that in the Near East. This Sudanic culture made several highly influential technological innovations including the production of ceramic pottery and cotton textiles that directly influenced the later civilizations. South of the Sahara it has been traditional to cook cereals, sorghum and millets as porridge rather than baking a loaf. The very remarkable origin of ceramic pottery two thousand years before they were first produced in the Near East enabled this kind of cookery, while the lack of ceramic pottery north of the Sahara encouraged the development of a different technology, baking, as a means of turning grain into a convenient food.

Sorghum is today the staple food for 500 million people in 30 countries in Africa, India and China. It matures in as little as 75 days and is highly photosynthetically efficient. It can grow in both temperate and tropical conditions but is especially favored in tropical conditions where maize cultivation is marginal because of drought, salt and water-logging. It is utilized in various ways; boiled like rice, cracked like oats for porridge, malted like barley for beer, baked into flatbreads, popped for snacks, eaten green like sweet corn. Nutritional quality as except it has a high tannin content that depresses nutrient absorption and a large quantity of protein content is the poorly absorbed prolamine. There are several different types of *Sorghum arundinaceum* including durra the main grain in Africa, milo the main grain in Central America, kaolang grown in China and feterita which has large red-yellow or white grains).

Further west in Africa in the region to the south of Lake MegaChad pearl or bulrush millet (*Pennisetum glaucum, P. americanum*) was taken into cultivation. It has a relatively low yield but is more adapted to heat and drought than maize. It is more nutritious than sorghum. Steam cooked it is couscous but it is also used to make breads, fermented foods and porridges. Fonio (Acha) *Digitaria exilis* (white fonio) and *Digitaria iburua* (black fonio) are two more kinds of millet grown in the West African savanna region. Fonio is favored because of its taste and is rich in methionine and cysteine, normally deficient in grains. It mature very rapidly 6-8 weeks after sowing at its quickest. Yet another millet, finger millet (*Eleusine coracana*) originated in central Africa in and around the highlands of Uganda. It is a useful grain because it is rich in the amino acid methionine lacking in the other tropical staples cassava and plantain. However it is a demanding crop requiring intensive cultivation at weeding and harvesting stages. It is interesting that the cultivation of sorghum, pearl and finger millets, and cotton were taken to India, at a very early stage, through Yemen, along the so-called Sabaean Lane, by-passing Egypt. In Asia new varieties of these crops arose.

There are many other millets some of which are used for grain, as livestock feed, but many only as forage grasses. They tend to be fast growing weedy plants. Indeed Kodo millet (*Paspalum scrobiculatum*), now grown mainly as a forage grass, though on the Deccan plateau of India it is a significant crop. It is regarded as an invasive noxious weed by the US Federal Government. Of the *Panicum* millets Prove, Proso or Common millet (*P. miliaceum*) has a long history of cultivation for its grain. It probably originated in central Asia but was cultivated by Neolithic European Lake Dwellers and was the *milium* of the Romans. It is used to make Russian porridge and is added to soups.

Oats *Avena sativa* also mainly have a reputation as the ingredient of porridge but also have been used to make oatcakes. They have the reputation as animal feed but are only behind wheat, maize and barley in importance as a cereal of temperate regions. The oat kernel is rich in protein and fat. Oats probably originated in central Europe about 1000 BC, perhaps first as a weed of wheat and barley fields. Several species remain as some of the worst weeds of these grains. Selection for a non-shattering inflorescence and lack of dormancy would have ensured its survival and propagation along with these crops. The husk contributes up to 25% of the weight and although a naked grained variant called Pilcorn evolved, it has since been lost, so that the overall useful yield of oats is significantly less than wheat and barley.

Grasslands are also very important as pasture for domesticated animals, especially cattle and sheep, and are also cultivated as fodder crops or for silage. Some pure grasslands like maize are cultivated but mixed grasslands, generally dominated by *Festuca* (Fescues) or *Lolium* (perennial ryegrass) are the most important source of fodder. Usually they are improved by mixture with legumes, like alfalfa and clover, which because of their nitrogen fixing abilities, produce a fodder or silage with higher protein content. Legumes are sometimes sown separately for fodder and various kinds of leafy vegetables (beet, swedes, carrots) are also sometimes used. In drier areas grassy sorghums, including Sudan grass (*Sorghum sudanense*) are widely grown as fodder and pasture.

There are a few other species that have been harvested as grains but are not grasses, especially from the family Polygonaceae. They were important in Neolithic cultures and have been important famine food but never quite made it as staple foods. Buckwheat

Fagopyrum esculentum and *F. tataricum* is the most important. It is used today mainly as food for poultry, buckwheat also provides a kind of flour useful for gluten free diets. In Central Asia and the Ukraine braised crushed buckwheat or kasha, is often combined with cracked wheat in recipes. Buckwheat flour is also used in the Far East to make noodles. Bees make a dark, highly flavored honey from the pollen and nectar. It produces the eight essential amino acids.

The role of the second and related family, the Amaranthaceae, in the emergence of cultivation in the Americas has not been emphasized enough. *Amaranthus* provides several grains and also leafy spinach-like vegetable. *A. caudatus* is Inca wheat. In North America different species *A. hypochondriacus* and *A. cruentus* were cultivated. The amaranths have the advantage of being better source of protein, they are rich in the amino acid lysine for example, than the cereals. They can be milled and "popped" like maize too. An additional advantage is their relative resistance to drought. All in all it's a pity that following European conquest they have been largely replaced by cereals. Some species, *A. albus*, *A. rudis* and *A. retroflexus* have got their own back; they are important weeds of a range of crops including corn, soya and sugar beet. Amaranths are also found outside the Americas and have formed part of the human diet, but here they have been mainly used as a leafy crop like spinach.

Quinoa (*Chenopodium quinoa*) is a similar "grain" popular especially with health-food faddists. It is in a third related family called the Chenopodiaceae. Cañahua is from the related species *C. pallidicaule* and there is a third called Huauzontle that is cultivated rarely in Mexico. It is still a significant grain in the Andes. Cultivation of quinoa emerged about 1500BC in the Andes. Fields at Chirapa in Bolivia dating from this age have a mixture of domesticated quinoa and its weedy relative, quinoa negra (*C. negra noa negra* (*C. quinoa* var. *melanospermum*). By 800 BC the fields are of pure domesticated variety. Quinoa flour can be made into biscuits, porridge or even bread. *Chenopodium album* Lamb's Quarters was an important forage food in Neolithic times.

Maize and the shaping of men

Quetzalcoatl, the plumed serpent, was the discoverer of maize in Central American mythology. He disguised himself as an ant to retrieve a single grain hidden inside a mountain by the ants. A range of plants were taken into cultivation in the Americas including avocado, squash, pumpkin, sunflower and amaranth but it is the cultivation of maize that led to the development of the great Central American civilizations. Between 5000 and 3500 B.C.E. maize and beans were also being domesticated though it has been calculated that such domesticated crops may only have provided about 10% of the diet. By the time of the Mayan civilization maize provided about 75% of the diet. In Mayan mythology the Creators took maize that grew out of the crack in the cosmic turtles shell and formed the shape of men, the ancestors of all Mayans. Its cultivation is associated with the rise of Mayan civilization and some Mayan babies had their foreheads bound with a board front and back to elongate them like a maize cob, and their hair cut to leave a tassel like the tuft on top of the cob.

The long period of maize domestication may be related to the extent that cultivated maize was transmogrified from its wild ancestor. The early genetic history of cultivated maize, *Zea mays* (from the Greek for corn *zeia*) is obscure but it is related to teosinte *Zea mexicana*, a weedy annual that looks, at least superficially very different, so different that for many years it was placed in its own genus called *Euchlaena*. Maize and teosinte have separate male and female inflorescences on the same plant but differ especially in the form of the female inflorescence. Teosinte has a series of lateral branches each with a terminal male tassel and lateral female spikes compared to maize with a strong main stem with a terminal male tassel and lateral female cobs. In teosinte the female spike is small and slender with 6-12 grains arranged alternately. Each grain is surrounded by a hard case. In maize in the cob there are several rows of grain each in a shallow soft cupule. However teosinte and maize hybridize easily and produce an intermediate hybrid plant. The origin of maize may have been by transmogrification of the corn so that each lateral branch was shortened and the terminal male tassel feminized.

In addition there was selection for greater size of cob with more rows of corn. In Teotihuacan of 5000 B.C.E. the average cob size was only about 2 cm but by 3400-2300 B.C.E. they were 4.3 cm, and by 700 C.E. they were reaching the average pre-conquest size of about 13 cm.

The earliest well-developed American civilization in Central America was that of the Olmec dating to about 1000 B.C.E. It originated on the coastal plain of the Gulf of Mexico, in a region of high annual rainfall, a period of lower winter rainfall allows the rivers to recede leaving rich alluvial soils. On these soils there was the potential of a regular bumper crop. On the more upland soils, more liable to nutrient depletion, nevertheless two crops were possible. Rich food surpluses gave rise to a stratified society with an elite military/religious caste controlling the richer lands. The eventual decline of this culture may have resulted from over-exploitation of the marginal slopes leading to nutrient exhaustion and soil erosion, as it was later to lead to the collapse of the Mayan empire. The lack of any significant technological development in agriculture, the absence of draught animals for ploughing or manuring make these Central American cultures distinct from those in the Old World.

Arising even earlier were the South American civilizations of coastal Peru. By about 4500 B.C.E a well-developed agricultural system cultivating maize, potatoes, quinoa, squash and beans was established. Agriculture was reliant on extensive irrigation works whose development led to a settled village life by about 2500 B.C.E. A rich highland Chavin culture that reached down into the coastal plain was contemporaneous with the Olmec but faded away about 300 B.C.E. In its place on the coastal lands the Mochica civilization arose about 200 C.E. but it too collapsed about 700 C.E. Climate change, increasing drought, made each successive coastal culture more and more reliant of extensive irrigation canals distributing water brought from the highlands for their agriculture, but inevitably exposed them to ultimate collapse in extended periods of drought. The last flourish of a coastal civilization was the Chimu, rising to prominence about 1200 C.E. with their city of Chan Chan which had a population of 25-50,000 people. But Chimu agriculture was also fragile, and they were easily conquered in 1465 when the Incas threatened to cut off their water supplies.

There has been very extensive diversification of cultivars in maize, like rice and wheat. Three major types of maize are grown; grain or field corn, sweetcorn and popcorn. Grain or field corn is by far the main kind; 70% fed to animals and more used as silage. It is classified into four main types; Dent, Flint, Flour and Waxy which differ in the distribution and type of starch in the kernel. Dent - hard starch at the sides and a soft type in the center of kernel, but shrinking at apex on drying to produce a (starch = 30% amylose, 70%amylopectin); Flint - hard starch layer entirely surrounding the outer part of the kernel (30% amylose, 70% amylopectin); Flour - very thin layer of hard starch and almost entirely soft starch (30% amylose, 70% amylopectin); Waxy - starch consists almost entirely of amylopectin. Sweetcorn contains a high proportion of sucrose in the kernel and popcorn a high proportion of hard starch, pops as water in the soft interior expands. Whole corn can be ground to provide cornmeal to make corn bread, tortillas and mush. Hominy or posole is produced by soaking corn in lime before grinding.

In the 19th and the first half of the 20th century there was a transformation of maize cultivars by the construction of inbred lines and hybrid varieties. Now nearly all the maize grown in the US and Canada is hybrid corn produced by crossing inbred lines. Crosses are ensured by de-tasseling (i.e. emasculating one line) or by the manipulation of male sterility genes. The utilization of a particular sterility genotype exposed maize to attack by the mildew *Helmonthosporum maydis*.

The king of vegetables

The potato is the king of vegetables second only to cereals in its importance to humans and so it is not out of place to describe it here in a chapter on cereals. *Solanum tuberosum* is only one of several hundred species in its genus. The tomato *Lycopersicon* is also closely related. An important step in the domestication of potato was the selection of mutants with alkaloid free tubers. The Andean civilizations that culminated with the Inca empire all relied upon it. Several different species were cultivated in pre-Inca times: *Pitiquiña* (*S. stenotomum*), *limeña* (*S. goniocalyx*), *phureja* (*S. phureja*), *andigena* (*S. andigena* – ancestor of *S. tuberosum*) and *chaucha* (a hybrid between *Pitiquiña* and *andigena*) are all still grown. By the time it was being cultivated by the Incas many varieties already existed. The Quechua language has a thousand terms to describe the varieties and features of potatoes. Potatoes were represented in art. Potato-shaped pots were made. Potato gods were worshipped and rituals performed to ensure fertility.

It was the potato that enabled the spread of the Inca empire. In places where wheat and maize could not grow they developed a system of regular crop rotation that kept their potato fields fertile. More importantly potato enabled the spread of the Inca empire, because in the high frosty altitudes it could be transformed into chuño, that could be stored and traded, and converted to a kind of flour. Freeze dried by being left to freeze in the dry upland climate of the Andes, there are two main kinds, a dark ordinary sort, and a white sort, called tunta in Bolivia, that is covered with straw to shield it from the sun. Each day, for four or five days, the tubers are trod by bare feet to squeeze the moisture out of them. Finally the ordinary dark chuño is dried and stored but the tunta is then soaked in a shallow pool for two months and then dried. It is snow-white and very light and can be ground into a flour. If you have never eaten c chuño imagine an instant mashed potato. The Incas received chuño in tribute storing it as security against time of famine. It was traded with the lowlands extensively, carried down on the back of llamas. They also treated cassava flour in a similar way to a limited extent but the Incas had several other root crops.

There are four other significant Andean tuber crops each in a distinct family Oca (*Oxalis tuberose* Oxalidaceae) is only second to potatoes as a vegetable in the Andean highlands. Ulluco (*Ullucus tuberosus* Basellaceae) is another Andean plant with potato like tubers. Fourth in importance as a tuber crop in the Andes is Mashua, Añu or Osañu (*Tropaeolum tuberosum* Tropaeolaceae). Arracacha (*Arracacia xanthorrhiza* Apiaceae) is a kind of Peruvian parsnip. The first three can also be dehydrated to make chuño. Unfortunately after the Spanish conquest chuño became subject to commercial exploitation, with Spanish traders obtaining a kind of monopoly over it so that native workers had to pay high prices for it. In the silver mines of Potosi the native slaves were fed on little else.

In 1537 a conquistador records the European discovery of

potatoes finding a store of "truffles" after raiding a South American village. The history of the introduction of the potato into Europe and then the rest of the world has been worked out in some detail. William Turner traveled widely in Europe but failed to mention the potato in his Herbal published between 1551 and 1562. According to some accounts John Hawkins introduced them to England in 1563. However the first written account is from 1596 when John Gerard listed in his "Catalogue" as growing in his garden in Holborn, a district of London. Gerard describes them in more detail in his own Herbal published in 1597, giving them a whole chapter to themselves. Unfortunately its rather obscure how he obtained them, though both Sir Francis Drake and Sir Walter Raleigh are mixed up in it.

It seems likely that Gerard was given a tuber by Harriot, the agent of Sir Walter Raleigh, who was picked up and carried home in 1586 by Drake from Raleigh's sponsored colony in Virginia (the Island of Roanoke). In Harriot's own account of the Virginia settlement he describes three Indian "root" crops, Kaishcupenauk (Jerusalem artichoke), Okeepenauk (*Pachyma fries* a truffle-like fungus) and Openauk (*Apios americana* or *A. tuberosa* called variously Indian potato, bog potato or groundnut). The last of these was a legume that produced strings of tubers on long stolons, and was a staple food for Indians in eastern North America, sometimes transplanted close to their campsites but never domesticated. It was this that Gerard seems to have mistaken for potato though it is clear what he grew in his garden probably was already on board Drake's ship when Harriot was picked up in Virginia. Drake had first become acquainted with the potato on the Island of Mocha off the coast of Chile in 1577, and valued it as a ships store. A report from Thomas Cavendish's expedition 10 years later records that the natives kept the tubers in cases of straw ready to give the Spanish as tribute.

Gerard makes reference to Clusius' knowledge of the potato, and it is in Clusius' *Historia* (*Rariorum plantarum historia*) published in 1601 that we get the first accurate account of the potato's origins as being from Quito in Peru. Clusius stayed with Drake in 1581 but evidently did not obtain potatoes from him but instead he writes that he obtained two tubers from Belgium in 1588, and that it was common and frequent in cultivation in parts of Italy where it was cooked with mutton, like turnips and carrots, though this may be a different confusion, because an Italian Flora of 1606 fails to mention it.

It seems that there were possibly multiple introductions. The confusion about its origins continued throughout the 17th century, even though Robert Morison professor of botany at Oxford from 1669, recorded that Banister, a missionary in North America hadn't seen it in either Virginia or any other American provinces. Nevertheless Salmon in his Herbal published in 1710 listed three potatoes, *Battata* or sweet potato, Virginian potato or *Pappas vel Battata Virginiana*, and most remarkably,

"Pappa seu Battata Anglicana seu Hiberniana the English or Irish potato, which grows in vast plenty in many of our English gardens so that the roots are sold by bushels in our London Market".

Evidently despite the confusion about its origins it was well-established by the end of the century.

Salmon's describes Virginia potatoes in a way that identifies them as probably "Andigena" potatoes. Introduced from the Andes remained a horticultural curiosity in Europe. They were adapted to the short days of their origin. Salmon's description of the Irish potato is much closer to modern cultivars and it seems even by this time selection had given rise to clones adapted to the long days of northern Europe. Potatoes remained for a long time a garden crop or cattle food in most of Europe but very swiftly became a staple crop in Ireland. At first potatoes were treated as a special food, valuable to give strength to old people. In 1651 John Bauhin describes the use of potato in the Americas as a substitute for bread and flour described potato. The spread of potatoes as a staple food was rather slow. They were adapted to the short days of their origin. Cultivation as a food delicacy first occurred in Spain but an important vent was the selection of clones adapted to the long days so that cultivation in northern Europe became feasible. By the 18th century potato had taken on the reputation of the sweet potato as an aphrodisiac, a reputation enhanced by its association with the production of very large families in Ireland.

Potatoes were introduced back into the Americas, first Bermuda in 1613, and in 1621 to Virginia, but more important was their introduction by Scottish and Irish immigrants in from the early 18th century onwards. In the 18th century Kings Frederick the Great of Prussia and William of Germany encouraged their cultivation there, and having been fed on them while a prisoner in Germany during the seven Years War, Parmentier encouraged their cultivation in France. New varieties had shorter stalks and stolons, fewer flowers, larger leaves and fewer larger tubers. Many new varieties were selected. An example of the kind of haphazard selection that took place was the discovery of the Jersey Royal early potato. In 1880 Hugh de la Haye, a Jersey farmer invited some friends for dinner and after the meal showed them two huge potatoes which he had been given. One had 15 'eyes' which he cut into separate pieces giving each one as a present. Planted out, the following spring, they produced a large and early crop with a high premium in the markets of Britain.

By legend potatoes were introduced to Ireland, by Walter Raleigh, and first grown in his garden at Youghal. Another legend is that they floated ashore from wrecks of the Spanish Armada! In whatever way they arrived, they were being cultivated as a field crop in County Wicklow in the 1640s. By the late 17th century their field cultivation had become widely established. Potatoes flourished in the wet cool climate; their needs were small and cultivation of a small patch on raised and manured lazy-beds could easily supply a family. The population of Ireland rose from 1.5 million in 1760 to 9 million in 1840. Over-reliance on the potato was fatal. Potato blight caused by the fungus *Phytophthora infestans* became prevalent final causing a full-scale epidemic in the early 1840s. The potato famine 1845-46 decimated the population. By death and emigration the population of Ireland fell by 2.5 million. The pattern for emigration was set and by World War 1 a further 5.5 million people emigrated to Britain and America.

Very rapidly wherever the potato grew to rival the cereals in importance, it absorbed the ritual and symbolic importance of the cereals it replaced. In Ireland seed potatoes were sprinkled with holy water at Easter time, and other rituals were adopted throughout the year. Elsewhere the potato was incorporated in to ancient pagan harvest rites. The transformation of potatoes into grains in disguise was complete!

Harvest Festivals

The cultivation of cereals entered human culture profoundly, not just because of their nutritional value. Cereals came to carry social, ritual and symbolic value. In Ancient Egyptian mythology Isis shows humans wheat and barley and Osiris the use of agricultural implements to cultivate them. Isis and Osiris are highest in the Egyptian pantheon second only to Amun. Isis the goddess of earth is whom Osiris visits annually as the god of the Nile. Min was the god of fertility and plants. In the spring harvest festival of Min, the harvesters pretended to weep at the destruction of spirit in the wheat.

In Sumerian legend Enlil the supreme Sumerian god of the earth, is the creator of agricultural tools and sets the lesser gods the task of cultivating the land. But after forty years they rebel and humans are made to do the work. His consort is Ninlil or Ansud a grain goddess, the daughter of Ninshebargunnu an ancient agricultural goddess and Haia the god of stores. Ashnan is the daughter of Enlil and Ninlil and another grain goddess. In the epic of Gilgamesh dated to six thousand years ago both wheat and barley are mentioned and the people of Uruk have stores of seven years' grain against famine. A Sumerian text from a cuneiform tablet from the late third or early second millennium BC "the farmer's instruction" records the advice of the god Ninurta, son of Enlil, god of war and vegetation, which provides the earliest surviving guide to the cultivation of barley. In the Epic of Gilgamesh, the oldest written tale, Gilgamesh gives an offering of flour to the gods.

Deep in the culture is the association of cereal cultivation with the cycle of death and rebirth. In western Europe the last sheaf of corn was plaited into a shape, sometimes a human figure, the corn dolly or corn mother. It was kept over winter and carried around the newly sown fields or ploughed into the first furrow or burnt as a sacrifice to ensure another bountiful harvest. A similar figure is made from rice stalks in South East Asia and of maize stalks in Peru. Wheat symbolised fertility and in Roman wedding ceremonies the bride carried a sheaf of wheat and wore a garland of wheat in her hair.

In the Eleusinian plain west of Athens, the most sacred rituals of ancient Greece were held annually to honour Demeter (known as Ceres in Roman mythology) and her daughter Persephone. Demeter is the goddess of agriculture and fertility, including human fertility and Persephone the goddess of grain. In a reflection of the Osiris myth like grain Persephone dies and goes to the underworld and then comes back to life. Hades snatched Persephone but Demeter refused to eat bringing on winter until Persephone was returned to her, so Zeus sent Hermes into the underworld to release her, but Hades tricked her into eating a pomegranate seeds, variously related as one seed for each winter month or season, so she would spend winter, with him.

The Eleusinian mysteries have a root in the rituals of Minoan and Mycenaean cultures, and possibly even further back to ancient Egypt. Similar festivals celebrating the grain goddess were widespread in the Mediterranean region. Another Greek festival was the harvest festival of Thesmosphoria. The Romans celebrated Cerelia, the Hebrews Sukkoth. Agriculture also evolved, and quite separately, in central Mexico and Asia, each under their own unique circumstances and based on different grain species, maize in Mexico and rice in south-eastern Asia. What they share with Near East agriculture is symbolic value of these foods and here too there were harvest festivals, in China called Chung Ch'ui where women bake moon cakes.

The Eleusinian ritual involved initiates the drinking of kind of beer called kykeon, mashed grain meal flavoured with pennyroyal mint. The beer was not strongly alcoholic. It makes one think of the merissa beer/porridge, made from sorghum that is the staple dish in parts of Sudan. However whether it was the alcoholic content or not it transformed the participant. Merissa is said to be non-alcoholic in the morning when it is first made but its alcoholic content steadily increases as fermentation takes place.

Perhaps the transforming effect of the beer was because of the presence of hallucinogenic water soluble alkaloids from ergot, the toxic lipid soluble alkaloids having been separated out in the process of mashing. Ergot is a fungus that commonly grows on rye and other cereal grains, especially in wet summers. Ergot alkaloids cause ergotism in humans: "The Devil's curse" or "St Anthony's fire", raving and twitching, abortion, even gangrene, behaviour that through the centuries have been associated with witches. Demeter(Ceres) is said to be teetotal because she didn't drink wine, but she is also associated with the poppy, another plant rich in alkaloids.

There is a complex pantheon of gods associated with grain cultivation and harvesting. Female figures include Demeter (Greek), Freya (Norse and Anglo-Saxon), Ziva (eastern Europe) and male figures John Barleycorn (British), Lugh (Celtic), Adonis (Greek), Cronus (Roman) The Greeks had gods to represent every stage of the process Eubolos (ploughing and sowing), Eunostos (storage), Karmenor (reaping), Triptolemos (threshing), Trokhilos (grinding husked grain), Hestia (baking) as well as those that guaranteed to maintain fertility of the soil. The Saxon holiday of Lammas celebrates the grain harvest. There will be more about John Barleycorn later. The central importance of wheat is shown by its place in the marriage ceremony. In Roman times the bride carried a sheaf of wheat and grains of wheat were thrown later to be replaced by the virginal white rice. In Elizabethan times the wheat was no longer thrown but baked into cakes, we of course have one big wedding cake.

Then there is the symbolic role of wheat or bread baked from it in many cultures, not least in Judaeo-Christianity: the harvest festival, "And thou shalt observe the feast of weeks, of the first fruits of wheat harvest, and the feast of ingathering at the year's ending." Exodus 322:22 or the last supper "And as they did eat, Jesus took bread, and blessed, and break it, and gave to them, and said, Take, eat: this is my body" Luke 22:19. Most telling perhaps is Jesus' assertion in John 6, "I am the bread of life; he who comes to me shall not hunger". In this ceremony the association between wheat and death and rebirth is explicit.

It is a great irony that Quetzalcoatl, the discoverer of maize and hence the founder of native civilization in the Americas also caused its downfall. He starts as a god of waters and then as a serpent that links earth and heaven. The Mayans called hum Kukulcan, the Incas Viracocha. He brought them all their laws and science. IIc was also the bearded white god who was expected to return and whom fatally the natives mistook for the conquering Spaniard. The gods of maize are depicted emerging from a crack in the earth, arms outstretched, a symbol of rebirth. Quetzalcoatl created humans with his own blood, and the release of blood is a common feature of rituals in the Americas: blood sacrifices were offered to the maize god both by the Mayans and Aztecs.

An earlier American "grain" crop also has a deep seated religious and ceremonial importance. It was supposed to confer supernatural powers and had an association like other cereals with death and rebirth. In Peru chichi beer is made from amaranth is drunk. Bright red, blood red, varieties have been selected for and the women apply the red color, obtained by boiling the fruits, to their cheeks and dance with bundles of red amaranth representing babies on their back. For the Aztecs it was associated with human sacrifice; the ground grain was mixed with honey and blood and shaped into little idols that were eaten. Quite independently in Ancient Greece amaranth's everlasting flower was associated with immortality and so it was used to decorate tombs.

The rather gruesome nature of staple foods in rituals is also emphasized by the representation of potatoes in pottery in the Andean and the associated coastal cultures is. Pots dating from the Pre-Chimu to Inca periods include representations of potatoes with large painted eyes. Some show twin tubers that may have been thought to have a magical quality. Some pots represent anthropomorphic tubers with the nobly parts of a potato as heads. In many the heads have deeply incised eyes and a mouth with the upper lip and nose cut back to show the teeth in a bloody grin. This gruesome representation hints at a brutal blood sacrificial rite turning the mouth of the human into a potato eye to encourage fertility of the tubers.

The cultivation and domestication of the three main staple foods, the grains wheat, maize, rice, and also of potatoes transformed human society. They could be stored and traded. Their cultivation allowed humans to expand their range. The size of human populations increased and humans started to live in permanent villages and towns. Food surpluses became wealth and societies became hierarchical and differentiated.

The cultivation of these crops required planning and organization. Humans looked to the future, they sought to secure themselves against famine. They became anxious and sought certainties in religion. They created mythologies which explained their world.

3 Cornucopia

"... he saw an orchard
Closed by a pale – four spacious acres planted
With trees in bloom or weighted down for picking:
Pear trees, pomegranates, brilliant apples,
Luscious figs, and olives ripe and dark.
Fruit never failed upon these trees: winter
And summertime they bore, for through the year
The breathing Westwind ripened all in turn –
So one pear came to prime, and then another,
And so with apples, figs, and the vine's fruit
Empurpled in the royal vineyard there.
Currants were dried at one end, on a platform
Bare to the sun, beyond the vintage arbours
And vats the vintners trod; while near at hand
Were new grapes barely formed as the green bloom fell,
Or half-ripe clusters, faintly colouring.
After the vines came rows of vegetables
of all kinds that flourish in every season"

These lines from Homer's, The Odyssey from the 8th century B.C.E. describe the lush plenty that Odysseus saw in the gardens of Alcinoüs on arrival in Sicily. The Neolithic revolution led to a massive increase in human populations and with the spread of agriculture it wasn't just cereals that were being domesticated. The staples were accompanied by fruits and vegetables of all sorts. The cornucopia symbolized the bounty that was available. It was the horn of a goat that suckled Zeus. Broken off it filled with fruit and grain, full of whatever was desired, a horn of plenty. All parts of plants have been utilized as a source of carbohydrates and other nutrients: from the root (root-tubers, taproots), stem (tubers, rhizomes, and canes), leaf, flower (nectar and pollen in honey), seed and fruit.

What is quite staggering is that it is only a tiny handful of plant species that have been domesticated; from a possible 250,000-300,000 species in the wild only about 2,000-3,000 species have been domesticated and only 100 of these are really important ones

(<0.04%). Of the thousands of species utilized by humanity only a relative few have entered world commerce. Less than twenty species provide most of the world's food. Why have these species become so important? Primitive cultures exploit a much wider range of wild species. Obviously the chosen crops had some features that favored them. However, there must have been a large element of chance in their selection. It is startling to think we might have ended up with a very different array of crops.

The transformation of domesticated plants

Some plants utilized by humans, especially the crops from trees and grasses, are still exploited directly from nature, natural populations still utilized as a source of timber or grazing and there the exploited plants are unchanged from wild plants, though the vegetation they are found in has in many cases been profoundly altered by human exploitation. Many of fruits and vegetables had been part of the earlier hunter-gatherer and vegecultures have been radically changed by their relationship with us and thoroughly domesticated. Once chosen they came under intense evolutionary pressure; selected for favorable traits they evolved rapidly. If early selection of improved varieties was accidental it soon became more conscious as particularly favored plants were tended and propagated. Some seed or roots from a particularly palatable plant or one particularly easy to harvest were saved and sown to provide more. Generation after generation of this kind of selection transformed the plant. Only a few are like maize, so transformed that they are quite unlike any possible wild ancestor, but many cannot maintain or propagate themselves with human intervention.

Changes to the domesticated plant were most extreme in the part chosen to provide food. Reproductive organs, seeds and fruits and other organs that like tubers that permit the plant to survive from one season to the next have often been chosen because they are storage organs, storing highly nutritious compounds proteins, lipids sugars, and especially starches.

As they have been domesticated their diversity has been exploited and their genetic pool fragmented into thousands of local cultivars or land races adapted to local conditions of climate and soil or local peculiarities of cultivation. Perhaps the chosen crop plants are species that have proved especially susceptible to domestication. They are the ones that have evolved most rapidly towards favorable characteristics for cultivation or for increased palatability. Genetic variation has been available or new mutations have arisen to be selected out. Hybridization between related species and existing cultivars has greatly expanded the genetic variation that is available for selection. The result has been an extraordinary proliferation of diverse vegetables from relatively few starting points, the ancestral crop species.

Crop plants share many features that differentiate them from their wild ancestors. Of course yield has been increased markedly but this has come about in several different ways: the size of the harvested organs and in some cases the number of harvested organs per plant has increased. Or there have been changes in branching pattern associated with changes in yield. Increased yield has been achieved in some cases by selecting a more favorable allocation of the energy budget of the plant, more favorable partition, towards the part of the plant that is harvested; the amount and size of grain produced relative to the foliage in cereals for example. The proportion of leaf to grain is smaller in dwarf plants and so being dwarf has been selected. Alternatively in fiber plants longer unbranched stems may have been selected, and the length and strength or flexibility of fibers has been improved.

A whole set of changes has improved the quality of the crop as a food; reduced spines and toxic constituents, increased sugar or starch content, increased attractiveness. Important adaptations have been related to agricultural practice. An inflorescence that does not break apart allows easier harvesting is one example. One of the most important changes in this century has been the selection of varieties that are able to make use of the high levels of fertilizers made available to them. Cultivars have become adapted to different climatic regimes and different photoperiods (that is, day-lengths in different latitudes).

In some respects, crop plants have been pushed towards behaving like a weed. They have faced less competition. Irrigation and the application of fertilizers have minimized stress. A rapid or annual life cycle has been selected. Smaller plants have been selected sometimes because they are associated with a shorter life cycle. Seeds and tubers lacking dormancy and with rapid and uniform germination have been selected. Synchronous development and ripening has been selected to enable more efficient harvesting. Cultivars have diminished ability to disperse compared to their progenitors.

Root systems, taproots or the number of tubers have expanded. The tubers, taproots and rhizomes of many such as potatoes and various kinds of yams have already been mentioned, but there are many others. Arrowroot *Maranta arundinacea* is mostly grown on St Vincent in the West Indies. It has cylindrical fleshy rhizomes, covered in the scars of leaf scales, that are ground up to make an easily digested starch, often used in baby foods. It has a strange history and seems only to have been domesticated until the 18th century and was previously used only as an antidote to the arrow poison manicheel from *Hippomane manchinella*. There is another New World arrowroot, *Canna edulis*, with a much longer history of domestication, called Achira, or confusingly Queensland Arrowroot where it has since been cultivated. Analysis of the starch grains found on grinding stones indicate that it was one of several tropical root crops utilized very early on, between 5000 and 3000 B.C.E. in lowland tropical America but it was later cultivated on the coastal plain of Peru and by the Incas in the Andes on the banks of their irrigation ditches.

Carrots (*Daucus carota*) are an example of selection for a more succulent taproot. Here selection has also changed the color; early domesticates in China included purple and yellow variants but in Europe the more appetizing yellow variants were selected and the color intensified, especially in the Netherlands about 1600 C.E. Another root crop from the same Apiaceae family, are parsnips (*Pastinaca sativa*). Before the introduction of potatoes both carrots and parsnips were very important root crops in the cooler parts of Europe and Asia. The wild ancestor of cultivated varieties are still common in the Northern hemisphere and both domesticated plants are close to their wild ancestors. Wild varieties are mainly biennial and delay the production of any taproot until the end of the first season. The cultivated forms behave more as annuals but have precocious taproot expansion; two years are squeezed into one and we harvest them before they flower.

In many crops there has been an alternate selection pathway, for succulence, softness of organs has been selected, by delaying reproduction, and by extending the life of soft juvenile organs. There are many kinds of legume sprouts and cresses (see below) but many other plants are used as seedlings or sprouts including clover, fenugreek, sesame, sunflower, wheat, rye and triticale. Almost any bamboo species can be eaten as bamboo shoots. Commonly cultivated ones are *Arundinaria* spp., *Bambusa beecheyana*, *Dendrocalamus latiflorus*, *Phyllostachys edulis*, *Sasa* species. Plants shoot in spring producing, large spring shoots 7-12 cm in diameter and up to 25 cm long and in the autumn producing the smaller "winter" shoots. Farmers tread the fields barefoot to feel the bumps that are the sign of a sprout, and heaping up the soil around it, or even placing a wooden box over it, to delay it becoming fibrous.

In some root crops, including turnips and swedes, it is the hypocotyl, the part of the plant below the seed leaves, the cotyledons, between stem and root system, that is expanded. In asparagus (*Asparagus officinalis*) it is the stem, or rather the young shoots that sprout from a dense rhizome system. Leaves are scale like anyway on the mature plant but selection has favored the production of large soft juvenile stems that can be harvested. By leaving a few steams to mature the rhizome system is enlarged to provide the storage organ that will feed the development of the next year's shoots. Plants are only male or only female (dioecious) and male only varieties are particular good because they do not waste energy in producing seed but put the energy into the rhizome system so that year by year fatter and fatter shoots are produced.

Leaves have been selected to be fleshier. Bulbs are tightly packed fleshy leaves were an obvious choice for domestication, especially in the onion family (see below). In celery (*Apium graveolens* var. *dulce*) it is the leaf stalks, the petioles, that have become fleshier, while in the closely related celeriac (*A. graveolens* var. *rapaceum*) it is the hypocotyl (the stem region below the leaves) that has become fleshy. The kind of selection for that has taken place for edible leaves is well-illustrated by the extraordinary varieties of lettuce (*Lactuca sativa*), broadly divided into cos, loose leaf and heading types but with a multiplicity of varieties of each. Pliny reports nine different varieties used in Rome in the 1st century C.E.

and Thomas Jefferson grew 19 varieties in his garden at Monticello in the 18th century. Relatively few are available in supermarkets. Lollo Rosso, or its green partner Lollo Verde are ubiquitous in mixed packs of salad leaves, dare one say, because their crispy serrated leaves bulk up any packet making it look full. In lettuce, as in other leaf crops, domestication has involved the suppression of the lateral buds and delaying stem extension so that a compact head is produced. In lettuce there has also been a reduction in the bitter milky sap, the Latin name *lactuca* refers to the milky sap, though at times this has been used medicinally as a sedative and mild narcotic.

The inflorescence, the flower buds have been selected too, notably in the brassicas. In the globe artichoke *Cynara scolymus* it is the bracts surrounding the inflorescence that have been selected. They have an ancient history of use. In Ancient Egypt they were associated with sacrifice and fertility. It has a reputation as an aid to digestion. The plant contains phenylpropanoids such as cynarin.

Any attempt to systematically categorize plants according to their uses is doomed to failure. Plants are a source of food and pleasure and material for construction. The same species can provide all three. Any further categorization of food plants becomes even more difficult. Is sugar from sugarcane a food or a pleasure? As rum it certainly is a pleasure to most but a necessity for some. What are fruits and what are vegetables? Bananas are fruits but plantains are vegetables and yet they are the same species *Musa acuminata*. The following is simply a convenient way of dividing up plants.

Some crop plants have a complex genetic history. There have been mutations with a profound effect on the morphology of the plant and also the gradual accumulation of genes with minor but cumulative effect. Hybridization has been an important source of new variation. Polyploidy is especially frequent. Some crops are self-pollinating derivatives of outbreeding ancestors. Inbreeding is widespread because it creates a homogeneous highly homozygous cultivar amenable to agriculture. However the crop may also be very vulnerable to disease. Some crops have become adapted for clonal reproduction at the expense of sexual reproduction. In other crops like maize techniques have been developed to produce hybrid varieties which are grown as highly heterozygous but homogeneous crops. Many of these genetic changes occur naturally in wild plants and have resulted in the patterns of diversity seen in nature. An interesting example is the mutation or series of mutations in maize that converted a male inflorescence into a female cob.

The families of crop plants

Some families have provided a large proportion of species exploited for food by humankind. In the forefront are the grasses (Poaceae) that provide the world with most of its carbohydrate either directly from the grain of the cereal crops (wheat, rice, maize, barley, oats, sorghum and millet) or as feed to animals. Cereals occupy over 70% of the world's croplands.

Other important families are the Fabaceae (legumes and beans of all sorts), the Solanaceae (fruits and the potato) and the Brassicaceae (leaf, root and oil-seed crops). In different parts of the world a different range of foods was domesticated but in each case as well as a cereal a legume a root crop was also cultivated. Other families important as sources of vegetables are the Alliaceae (onions, leeks, garlic, shallots), and Cucurbitaceae (cucumber, gherkins, squash, marrow and pumpkin).

Many crop species have multiple uses. They were initially domesticated for one purpose and then a different part of the plant was exploited. There are many different fruits from the Solanaceae that are also treated as vegetables. The Solanaceae, as well as providing potatoes, provided a range of peppers and tomatoes. Chilies have been used for at least 7000 years to flavor dishes in Central America. Chili peppers are a variety of *C. annuum* that also provides sweet peppers and paprika.

Tomatoes come 16 in the top 30 world crops. Treated as a vegetable they are of course a fruit. In some dictionary definitions a vegetable is a plant eaten with an entrée or main course or in a salad but not as a dessert or table fruit. In fact in 1893 the United States Supreme Court ruled that a tomato was legally a vegetable rather than a fruit. By this definition a vegetable may originate from any part if the plant, including the flower as in the globe artichoke (*Cynara scolymus*).

Tomatoes and bell and chili peppers have an American origin; they were widely cultivated in the Americas before the arrival of Columbus. There are also other less well known fruits from the same family such as the tree tomato (*Solanum betaceum*), naranjilla or lulo (*Solanum quitoense*) and pepino (*Solanum muricatum*) from various parts of the Andes and others elsewhere.

Some fruit crops in the same family have an Old World origin. Aubergine (eggplant or brinjal) *Solanum melongena* are the most important, indeed after tomatoes and potatoes they are the most important crop in the Solanaceae, holding the position as the king of vegetables in China and India over its rivals. *S. melongena* was probably first domesticated in India where there is a wild ancestor which is perennial and has spiny fruit, and there are four other related species. From there it seems to have spread to southeast China and the Middle East. It was the Arabs who introduced them to Europe through Spain and Sicily. Caponata is an ancient Sicilian aubergine dish probably established during the Arab occupation of Sicily. The name aubergine comes from the Catalan *albergínia* from the Arabic *Al-badhinjan*. Moussaka is a Balkan dish of aubergines with a Turkish and Arabic origin. The genetic ancestry of aubergines is complex. Different kinds of aubergine, more bitter and usually yellow, were domesticated in Africa *Solanum aethiopicum*, jaxatu is a scarlet African eggplant, often eaten before it is ripe. *S. macrocarpon* is another African eggplant called gboma.

In the Americas the squashes Cucurbitaceae were important sources of vegetable fruits. In Mexico three main kinds of squash were domesticated; warty or crookneck squash (*C. mixta*), walnut squash (*C. moschata*) and pumpkin *Cucurbita pepo*. The terminology is complicated because there are about 26 species of Cucurbita which hybridize. Other important domesticated species *C. maxima*, and *C. ficifolia*. The domesticated species lack the thick rind and bitter flesh of wild relatives. Each species has been selected to provide different products. *C. pepo* provides marrow, zucchini or courgette and gem squash. The date of domestication is very early. Domesticated *C. pepo* has been dated to 8000-7000 BC at Guilá Naquitz so it may be the earliest plant to be domesticated in the Americas.

Potage and pudding

Pease pudding hot, pease pudding cold,
Pease pudding in the pot – nine days old.
Some like it hot, some like it cold,

Some like it in the pot –nine days old.

Pease pudding is only one of many puddings, potages or porridges made with a variety of cereals and pulses. Barley, wheat but also oats and rye were all utilized along with peas, beans, lentils and flavored in a variety of ways – a lump of boiled ham in the case of pease pudding. Theophrastus, the father of botany treated the pulses as cereals in his *De historia plantarum*. A favorite dish of the Roman emperor Augustus was *tisanum*, a porridge of mixed pulses; chickpeas, lentils and peas with crushed barley.

The pulses are all legumes from the family Fabaceae (the old family name is Leguminosae). There are many different species. Beans (Adzuki, Anasazi, Appaloosa, Black, Black-eye, Cannellini, Chickpeas, Fava or Faba, Great Northern, Jacob's Cattle, Kidney, Lentils, Lima, Navy, Peas, Pinto, Swedish Brown, Yellow-eye). In some ways beans are surrogate cereals. Lentils and pulses have provided an important food that could be preserved dried, another surrogate grain, they were one of the staples in the Near East eight thousand years ago. Jacob traded lentils to Esau for his birthright and are part of a bread made by the Jews in Babylonian captivity. Humus and dahl are two dishes we are happy to eat but how many people still eat our own pease pudding, once as much a staple of our diet as porridge.

Legumes are especially important as food not only because they are 2-3x richer in protein than cereals but also because the spectrum of proteins they contain is different from cereals. They are especially important as a source of protein for poorer people in the tropics. Soybeans (*Glycine max*) are said to contain all the essential amino acids in a quantity sufficient for an adult. Legumes are generally rich in lysine that is in low concentration in cereal protein, but they are generally low in methionine and cysteine. The range of proteins is very great though this can cause problem. Peanut allergy, mainly to a number of low molecular weight proteins in peanut, has become a serious problem in the west in recent decades.

Faba or fava beans have been cultivated for many millennia in the Mediterranean region even though some classical authors have advised against eating them. In Greece they were associated with funeral ritual; even the great Pythagoras says that the hollow stalk of the plant provides a route to Hades. Perhaps this relates to the fact that a large minority of people in this region have favism, an allergic sensitivity to eating the beans or even inhaling their pollen, causing hemolytic anemia. People with favism have a form of genetic deficiency in an enzyme called glucose-6-phosphate dehydrogenase(G-6-PD). This is the commonest human enzyme deficiency whose distribution is associated with the distribution of malaria in past times. Its distribution is associated with the disease particularly affects males. Nevertheless faba beans were a favorite Roman vegetable. Maccu, a soup made with mashed beans, flavored with wild fennel, is even today Sicilian peasant food.

Legumes are also an important source of dietary fiber. Legumes differ in their digestibility. As well as being rich in proteins pulses are often rich in highly fermentable compounds, such as soluble dietary fiber and resistant starch. People vary in their ability to digest beans, as we all know, causing terrible flatulence in some cases. Indigestible short chains of sugars, oligosaccharides, have either 5 sugars (verbascose), 4-sugars (stachyose) or 3-sugars (raffinose). Stachyose is the main oligosaccharide in pulses and they also contain raffinose (onions and chicory, and brassica foods like broccoli are higher in raffinose than stachyose). These oligosaccharides are sometimes known as raffinose-series oligosaccharides (RSO) or alpha-galacto-oligosaccharides because of the way the sugars are linked. Animals with a single stomach like humans, pigs and poultry, do not have alpha-galactosidase enzymes to break them down and they pass through to the lower bowel, the caecum and large intestine, where bacteria ferment them producing carbon dioxide, methane and hydrogen. Apparently only a third of people produce methane, potentially inflammable farts. Cooking reduces the problem, halving the raffinose and stachyose content of common beans for example. Animal feeds, like soya, that contain these components are treated before use to prevent problems to farm animals. One method of treatment is to allow a natural fermentation to take place before they are ingested.

Another way of reducing oligosaccharide content is to allow the seed to germinate and to eat the seedling or shoots. Mung bean sprouts (*Phaseolus aureus*) have commonly been used in Chinese dishes but have escaped into a wide range of salads. Several other kinds of beans are often used including Soya, adzuki, and garbanzo beans and even lentils, green and chick peas and alfalfa (*Medicago sativa*).

Lentils, peas, vetches and beans were some of the earliest crop plants to be cultivated. In the lupin it has been possible to observe the process of domestication because it has taken place recently. *Lupinus albus* were eaten by the Romans, but were mainly used as fodder. However in the 20th century more determined efforts were made to domesticate other species of lupin. Mutants with lower alkaloid content (more sweet), reduced pod shattering, a more permeable seed coat, early flowering and disease resistance were all relatively quickly selected.

The common bean *Phaseolus vulgaris*, was one of the trinity of domesticated plants upon which American civilization was founded along with maize and squash. Mixed fields had advantages; the squash could climb up the maize stems and crucially the beans enriched the nitrogen content of the soil with their nodulated roots inhabited by atmospheric nitrogen fixing *Rhizobium* bacteria. Common beans have been detected in early archaeological deposits, pre-dated only by squash as a domesticate, from 8000 B.C.E. though at this time it was probably being grown as a green vegetable for its tender shoots and pods. Domesticated beans don't appear until 4000-3000 B.C.E. by which time they had increased seed-coat permeability to allow quicker soaking before cooking, a pod that was no longer brittle or corkscrew shaped so that it could be harvested an shelled easier and in some varieties a shift to an annual habit. *Phaseolus* may also have been domesticated in Brazil and northern Argentina. Now it provides common, haricot, navy and snap bean. Other important bean species are the closely related *P. coccineus* or runner bean and *P. acutifolius* or tepary bean, and the more distantly related *P. lunatus* or Lima bean. The latter seems also to have been domesticated in South as well as Central America.

In South America, another legume, the peanut was being cultivated from at least 6000 B.C.E., from which time remains have been found in the Zana valley of northern Peru. The peanut is actually a seed is the peanut; the monkey nut is the legume of *Arachis hypogaea* and the peanuts the two seeds it contains. It had probably been cultivated for much longer because *Arachis hypogaea* is a hybrid species with double the number of chromosomes (allopolyploid) of the parental species from which it is derived. The parental species that are now only found on the eastern slopes of the

Andes; the center of diversity of the genus is in Bolivia. Peanut was still being cultivated in the coastal plain of Peru the time of European arrival and from there introduced to Europe and thence North America and Africa.

In Africa it has diversified again. It is now the most widely cultivated legume and one of the five most important sources of vegetable oil. The seed has a 20% protein content and 50% oil content, 80% of which is unsaturated oils; in contrast most other legumes, with the exception of soya beans have seeds with less than 5% oil or fat. *Arachis hypogea* is interesting because of the way the stalk, the pedicel, elongates and becomes negatively geotropic after the flowers have been pollinated in order to plant the legume that contains the one or two seeds.

Many pulses have multiple uses. For example soya bean provides oil, fodder, flour used in confectionary, for biscuits and ice-cream, and in fermented form soy sauce. Soy milk is made by grinding and soaking the beans and straining the grounds out. Tofu is made by curdling soymilk with a coagulant like magnesium chloride (called nigari in Japan), the last component to precipitate out in evaporating sea-water, or calcium sulfate (gypsum).

Many legumes are also important fodder crops and green vegetables and leguminous trees are important sources of timber and fuel in the tropics.

Brassicas - exploiting all parts of the plant

The family Brassicaceae provide a remarkable example of a group of closely related species providing root, leaf and seed crops. The brassicas are out-standing example of a kind of plant in which almost all parts of the plant have been exploited in one way or another: hypocotyl – turnip and swede; stem – kohlrabi; shoot buds – Brussels sprouts; leaves – cabbage; petiole and leaf veins - pak choi, inflorescence head – cauliflower, plus flora buds – broccoli, cauliflower; seeds – rape. The seedlings and young plants provide cresses of all sorts. "Mustard and cress" is a mixture of two brassica species, white or yellow mustard, *Brassica hirta,* that used to be called *Sinapis alba*, and cress, *Lepidium sativum.* The cress takes 3-4 days longer to germinate. The mustard oils in the mustard provide a tasty hint of pungency. Water cress is *Nasturtium officinale* is another brassica and a crop that is identical in the wild and cultivation. However most other brassica vegetables have been astonishingly transformed by domestication and selection., an some species have provide a vast diversity of different vegetables.

Brassica oleracea, B. campestris, and *B. napus* are closely related species that exist as an astonishing range of crops: cabbages, kales, mustards, broccolis, cauliflowers, turnips, swedes, kohlrabi, sarson, and oilseed rapes. The Chinese and Japanese have their own range of brassicas for use as vegetables or salad plants. Most of them are varieties of *B. napus* (wong nga baak, baak choi, choi sum, mizuna etc.). The classification of the brassicas is exceedingly complex with the same variant recognized as a subspecies or a varieties or as a member of a subgroup or as a cultivar or by some other name!

Part of the difficulty is that *Brassica* species have hybridized to create new hybrid species that have further diversified. *B. napus* (swede) is a hybrid derived from a cross between *B. oleracaea* (cabbage) and *B. rapa* (turnip) that has doubled its number of chromosomes (an allotetraploid). Also called rutabaga from the Swedish rutabaga swede was first called in English speaking countries Swede turnip and in France navet de suede or chou de suede but this may not indicate an origin there – just that its cultivation became widespread there first. Its cultivation seems to have spread out of Sweden significantly in the 17th century.

The picture is even more complicated because hybridization between either cabbage and turnip and black mustard *B. nigra* has given rise to two more allotetraploid hybrid species *B. carinata* Ethiopian mustard and *B. juncea* (brown mustard. It doesn't even stop there because an older hybridization between *B. oleracea/rapa* and *B. nigra* gave rise to radish *Raphanus sativus* and white mustard *Sinapis alba* is also related. The whole group has been called a coenospecies from *coeno* meaning "common".

It gets even more interesting because it now seems that after the ancestor of the whole group split from a different genus called *Arabidopsis*, hybridization between them resulted in an ancestor with three times the chromosome number (hexaploid). *Arabidopsis* has been the subject of intense genetical study: it is the experimental *Drosophila* , the fruit-fly, of the botanical world, partly because it has relatively little DNA in comparison to other plants, and reproduces very rapidly. It has only 6 chromosomes and the *Brassica* group have 18 or multiples of 18 chromosomes[i]. It seems all this DNA, initially three copies of every gene that was present in Arabidopsis, has allowed the *Brassica* coenospecies to diversify in this very remarkable way.

Any one of these species, either hybrid or parental has numerous highly distinct variants. For example B. oleracea includes numerous varieties of cabbages, leafy kales, cauliflowers, broccolis, Brussel Sprouts, Kohl Rabis, Marrow-stem Kales. Look in any horticultural seed list and you will find many of the hundreds of varieties of cabbages "early-", "late-", "white", "savoy" – the list seems endless, and that is just cabbages – all just one botanical variety because they are so closely related.

Oil-seeds

These days one of the most important brassicas is oil seed rape that seems to fill our fields with yellow in spring. It has however been cultivated in Northern Europe. Oil seed crops are derived from a remarkably wide range of plant families. Oil palms have been mentioned already and the most important by far is Soybean.

The breeding and cultivation of oil-seed plants provides an outstanding example of the potential of even established crop plants to provide different needs. The varieties of turnip and swede that that oil seed rape cultivars are derived from has been cultivated since the Middle Ages, mainly as animal feed, and were used for example, in reclamation of polders, fenland and salt-marshes. *Brassica napus* (swede) and *B. rapa* (turnip) are relatives of the cabbage but have been bred for high oil content in their seed.

Many different varieties are grown for human consumption, in cooking oil and margarine, animal feed, or industrial use. Varieties cultivated for consumption have low erucic acid (<1%) and the sulfur containing compound glucosinolate content. Both these compounds reduce the digestibility and palatability of the crop. One disadvantage that low glucosinolate varieties have is that they are more susceptible to pests. Double low varieties are also marketed under the name "Canola". Alternatively HEAR varieties have high erucic acid content (50-60% of the oil). The erucic acid is converted to erucamide which used in the manufacture of polythene to reduce friction and prevent films sticking together. Manipulation of the fatty-acid spectrum and content provides oil for a multiplicity of purposes from detergents to pharmaceutical manufacture. For example a reduction in longer chain fatty acids such as linolenic acid content increases shelf-life for food-oils. Some manufacturing processes prefer high oleic acid content and high oleic acid content is also favored in the diet for the prevention of coronary heart disease. The potential market of rape seed oil is great. Already it is incorporated into lubricants for two stroke petrol engines but derived methyl esters could be used as a diesel substitute when mineral-derived diesel oil becomes more expensive.

Related to oil-seed rape are the mustard oils *B. juncea* and Sinapis alba cultivated in India and elsewhere for their oils but also for the production of the condiment. For this purpose it is the high glucosinolate content which provides the taste. *B. juncea* contributes volatile, or nasal pungency, and *S. alba* heat and sweetness in the mouth. B. nigra black mustard is one of the oldest recorded spices domesticated in Asia Minor or Iran, but in recent decades has been largely replaced by *B. juncea* which has better harvesting qualities (a non-dehiscent fruit (a silique)).

In lower latitudes than where oil-seed rape is normally cultivated sunflower (*Helianthus annua*) becomes the most significant oil seed crop. The oil has over 90% of oleic and linoleic acids in roughly equal proportions. As well as providing a cooking oil valued for the high linoleic acid content, it is useful in manufacture as a "drying oil" in paints and varnishes. It also does not yellow over time, in contrast to the nearly 50 % linolenic acid provided by linseed. It is the linolenic acid in some oils that leaves a pervasive odor when it is used for frying. Safflower (*Carthamnus tinctorius*) is another oil-seed from the Asteraceae.

Castor-oil, *Ricinis communis*, has little use in the diet, its value as a purgative is well-known. However it has extensive industrial use as a non-drying oil and was used in lamps by the Egyptians more than 4,000 years ago. Tallow tree Seeds, Karite Nuts and Tung Nuts are all valuable sources of oils and fats. Tallow tree, cultivated in provides fats for candlewax and soap. *Aleurites* species (Tung Nuts) cultivated in South East Asia provide drying oils for paints and varnishes, candle-nut oil and are also used in curries. Karite nuts are primarily grown in the semi-arid Sahel belt of Africa. A solid fat (butter or stearin) and the liquid oil (olein) have a wide range of uses, especially as a substitute for cocoa butter.

Onions and garlic, for strength and health

Vegetables in the onion family Alliaceae formed an important part of the diet in classical times. They were well known to the Ancient Egyptians and the Ancient Indians. Garlic, suan tai, was a staple of Chinese cuisine by 4000 years ago, along with spring onions, Ts'ung, and ginger, Geung, providing the basis of many stir fries. The wild ancestor of garlic is *A. longicuspis* which is native to SW Asia. Onions are a staple of the Mediterranean diet. A merchant from Baghdad visiting Palermo in Sicily in 977 C.E. disparaged how the locals ate large quantities of raw onions on a daily basis.

There are hundreds of species in the onion family and numerous food plants not just onion *Allium cepa*. The taxonomy is rather tangled because the species are often closely related and different vegetables are recognized either as varieties or species. Shallots are called *A. ascalonicum* or *A. cepa* var. *ascalonicum* or sometimes var. *aggregatum* though the latter varietal name is also used for spring onions. Shallots are said to have been introduced to Europe by returning crusaders from the Palestinian city of Ascalon , hence the name scallions applied to them, but also to spring onions or other immature green onions. Leeks are *Allium ampeloprasum* var. *porrum* or *A. porrum*, from the Latin for potage, porridge or soup. In Roman cuisine vegetables of the onion family were commonly pounded in a mortar before being added to a dish. Vernacular names only add to the confusion. Elephant Garlic is *A. ampeloprasum* var. *ampeloprasum* and is related to leek but most other garlics are different varieties of *A. sativum*. Soft-necked garlic is either called *Allium sativum* or *A. longicuspis* and hard-necked garlic *A. sativum ophioscordon* or *A. ophioscordon*. The latter has the cloves in a ring around a solid flowering stem. Other garlics include *Allium scorodoprasum* Rocambole, or Serpent Garlic with its twisted stems, and *Allium tuberosum* Garlic Chives or Chinese Chives.

Different parts of the plant are eaten. In chives (*A.*

schoenoprasum) and Japanese bunching onion (*A. fistulosum*) only the leaves are eaten. Siberian Chives or Chi si shan jiu are *Allium nutans*. In leeks (*A. ampeloprasum*) the basal part of the leaves as well as the bulb is eaten. Leeks were an important food in the Middle Ages because of their winter hardiness. Kurrat is an Egyptian and Near East version in which only the green part of the leaf is eaten. Rakkyo, (*A. chinense*) bulbs are usually pickled or canned like pickled onions. In both onions and garlic, the bulb is eaten though in garlic it is the expanded fleshy center, the "clove", not the surrounding flesh bulb leaves. In Chinese cuisine different varieties of garlic chives Gau choi are sliced for stir fries, either the flat green leaves , either green or yellow from being grown in the dark, or for the cylindrical flowering stem.

Garlic and rakkyo are unusual in that unlike onions and leeks they are not propagated by seeds but instead only asexually from bulb cloves. Garlic is viviparous and can also be propagated by bulbils. Bulbils are produced in the flower head in *Allium cepa proliferum* the Tree Onion and also in *Allium vineale* Crow Garlic.

In Chinese cuisine garlic is a hot herb, giving strength. Its medicinal value has been recognized for millennia. The pyramid builders in Egypt were given garlic to give them stamina, and even paid partly in garlic. Roman legionnaires ate garlic for strength. In India widows are not allowed to eat garlic because it is an aphrodisiac while in the west it enflames the lusts of men. At some point garlic became identified with keeping away evil. In Homer's Odyssey, Circe gives Odysseus a potion consisting of garlic that prevents him turning into a pig. It is woven into amulets in the Middle East, and in Europe hung over the entry to ward off evil spirits like vampires.

The medicinal use of garlic is ancient. Dioscorides mentions not just garlic but "*ophioscordon*" for dog and snake bite and as a laxative and vermifuge. Louis Pasteur described garlic's antimicrobial properties and that's not all; they also have antifungal, antiprotozoal, antiviral and anti-parasitic activity. It is also a prophylactic against some kinds of cancer probably because it is an immune-stimulant. Several large scale surveys have showed that regions in which large quantities of garlic are consumed have significantly less cancers of the gut. In one trial a group that took one garlic capsule each day had significantly less colds, and had less severe symptoms if they did catch cold, than a group taking a placebo. In recent years it has also been taken in order to lower cholesterol levels, and to thin the blood, although experimental trials have shown only a relatively small effect compared to prescription drugs or aspirin. *Allium ursinum* Ramsons, Wild Garlic or Bear's Garlic has a traditional herbal use for a variety of similar condition but has not been used much in cuisine.

The compound that gives garlic its strong smell is a sulfur compound called allicin (diallyl disulfide) that is formed in the air when garlic is crushed. An odorless compound called alliin is converted to 2-propenesulfenic acid by the enzyme alliinase and in the air this dimerises to form allicin (2-propenyl-2-propene thiosulfinate). Alliin is related to the amino acid cysteine. The bad breath and body odor of the garlic eater is caused by a cascade of sulfur containing compounds that are released when the thiosulfinates react with oxygen.

After harvesting garlics are dried and in this state can last for many months before sprouting, but eventually they will the green shoot trying to find their way off a fridge shelf: as they prepare for sprouting the flavor becomes more acrid. Commercial preparations of garlic that are odorless have questionable benefits since it is the smelly allicin which is the active component. In a similar way onions make you cry because when alliinase is released amino acid sulfoxides that in air breakdown forming propathenial S-oxide (sulfenic acid) that volatilizes. On meeting the water on your eyes these convert to sulfuric acid. The sulfenic acids convert to the onion smell of thiosulfinates when cooked. Shallots have a milder flavor.

The orchard

"And God said ...every tree, in which is the fruit of a tree yielding seed to you it shall be meat." Genesis 1:29.

"I Blesse the Lord, because I grow
Among thy trees, which in row
To thee both fruit and order ow," George Herbert

George Herbert's lines are part of a long tradition comparing the Garden of Eden to an orchard or botanic garden as if the trees in the

Garden were ordered in rows and tended by the divine hand. John Parkinson in his *"Paradisi in sole Paradisus terrestris"* of 1629, the first work in English to describe garden and orchard plants separately, has a famous title page illustrating the garden of Eden. It contains no animals except for Adam and Eve, and a strange vegetable sheep, the Scythian lamb, (vegetable lamb of Tartary). The lamb was actually the white fluffy trunk of the fern *Cibotium barometz*. In the centre of the garden was the tree of knowledge of good and evil looking like an apple tree. In Parkinson's book there is a chapter on apples and here he lists the main varieties of the time

"The Summer pippin is a very good apple first ripe, and therefore to be first spent, because it will not abide so long as the other.

The Golding pippin is the greatest and best of all sorts of pippins.

The spotted pippin is the most durable pippin of all other sorts.

The great pearmaine differeth little either in taste or durability from the pippin, and therefore next unto it is accounted the best of all apples.

The summer pearmaine is of equal goodness with the former, or rather a little more pleasing, especially for the time of its eating, which will not bee so long lasting, but it is spent and gone when the other beginneth to be good to eate.

The Broading is a very good apple.

The Flower of Kent is a faire yellowish greene apple both good and great.

The Gilloflower is a fine apple, and finely spotted.

The Gruntlin is somewhat a long apple, smaller at the crowne than at the stalke, and a reasonable good apple.

The gray Costard is a good great apple, somewhat whitish on the outside, and abideth the winter.

The Belle boon of two sorts winter and summer, both of them good apples, and fair fruit to look on, being yellow and of meane bigness.

The Dousan or apple John is a delicate fine fruit, well relished when it beginneth to be fit to be eaten, and endureth good longer than any other apple.

The Spicing is a well tasted fruite.

The Queen apple is of two sorts, both of them great faire apples, and well relished, but the greater is the best.

The Pot apple is a plaine Country apple.

The Cowsnout is no very good fruit.
The Cats head apple tooke the name of the likenesse, and is a reasonable good apple and great.
The Kentish Codlin is a fiare great greenish apple, very good to eate when it is ripe, but the best to coddle of all other apples.
The Geneting apple is a very pleasant and good apple.
The old wife is a very good, and well relished apple. The towne Crab is an hard apple , not so good to be eaten rawe as roasted, but is excellent to make Cider.
The Sugar apple is so called of the sweetnesse.
Sops in wine is so named both of the pleasantness of the fruit, and beautie of the apple.
Twenty sorts of Sweetings and none good."

A mouth-watering collection of names, and as F. Greenoak has pointed out almost a hymn to the variety of apples. Of all the plant species utilised by humans it is in the fruits that we come closest to sampling the wild abandoned diversity of nature. And occasionally we get the delicious pleasure of actually collecting our own fruits from nature. Black-berrying was my favourite as a child, poly bags heavy with fruit, covered in scratches but satisfied and looking forward to a delicious apple and blackberry crumble. Blackberries, *Rubus fruticosus*, is one of very few fruits that is still a truly wild harvest in Britain. Selection has changed most fruits, domesticated them so that they rely on human propagation to survive. Even though their relationship to their wild ancestors is often still very close, compared with their wild ancestors cultivated fruits are fleshier, sweeter, with fewer tannins and glycosides, and are frequently less seedy. Even in the tropics favoured trees are grown in a mixed "orchard" around the domicile and those still in the forest are tended and propagated.

For example in Turkestan there is a range of wild varieties of the apple (*Malus pumila*) that are large and sweet bridging the gap between the crab apple (*M. sylvestris*) and the domesticated varieties (*M. domestica*). The number of varieties is now huge. Pears *Pyrus* is a related genus that has a similar geographical origin. They have the same basic chromosome number (17) and the same kind of fruit, called technically a pome. Pliny records 22 varieties of apples. Pomme d'Api and Court Pendu Plat may be surviving Roman varieties. Old Pearmain is perhaps the oldest surviving English apple variety and dates back to the 1200s. Pearmain refers to the long or pear shape. Costard apples, large ribbed varieties, are almost as old. The latter were sold from street barrows and gives us the word costermonger. In England a distinction arose not just between dessert apples and cider apples, but between sweet dessert varieties for eating and large hard acidic varieties for cooking. The cookers hold their shape but soften in cooking as the acids are released. The particular flavor of ciders comes from the tannins present. Pippin, apple varieties raised from pips, (pepin in French) were probably introduced to Britain from the late Middle Ages. Cox's Orange Pippin was raised from a pip of the Ribston Pippin about 1825 by Richard Cox a retired brewer from Colnbrook in Buckinghamshire. Apple orchards established by Johnny Appleseed and others in

America gave rise to many varieties that have since become worldwide favorites such as Golden Delicious and Jonathan. The story of Johnny Appleseed, is like a legend. He was a nursery man called John Chapman born in 1774 in Massachusetts and one of the first to explore the new lands south of the Great Lakes that were then opening up in North America. For fifty years he lived like a hunter-gatherer though he ate no meat. Everywhere he traveled he carried a bag of apple seed and established little apple nurseries. It was a breeding experiment on a tremendous scale. New varieties came in to being, selected for the beauty and sweetness of their fruit, their resistance to diseases and their adaptation to their local conditions. By the time the settlers arrived there were young apple trees to sell.

Apples belong to one of two plant families that provides the majority of commercial fruit the Rosaceae is one of The fruits of the Rosaceae are very diverse including pomes (apple, pear, medlar, quince, azarole) drupes (peach, apricot, plum and cherry), drupecetum (= an aggregate of drupelets) (blackberry and raspberry) and pseudocarp (strawberry).

In contrast in the other commercially important family, the citrus family Rutaceae, the fruits are all of a single characteristic form, a kind of berry called a hesperidium with a succulent glandular hairy endocarp, but known to us as oranges, lemons, limes, grapefruit, pomelo(pummelo), tangerine, satsuma, and mandarin. They may vary in size from limes a few centimeters across to pomelos exceeding 30 cm.

It is in the fruits that we are most aware of the range of qualities of individual crop varieties. There are hundreds and hundreds of varieties of apples and citrus. As well as soft fruits there are many species harvested for nuts. In fact some of the most popular nuts like Brazil nuts and almonds are not nuts. Coconuts, almonds, and walnuts are the inner part of a drupe. Brazil nuts and kola nuts are seeds. The cashew nut is produced on the end of a fleshy aril (cashew-apple) that is also eaten and in Brazil made into a drink called cajuado. The nut is very hard and has to be cracked open by large lever like pliers, a process made more difficult by the acrid liquid that surrounds the seed. Each seed is individually poked out with a pin.

Some fruits, such as tomatoes and plantains are treated as vegetables. Indeed the tomato is deemed a vegetable by a ruling of the Supreme Court of the USA! The range of fruits eaten is astonishing. Today a wide range of tropical fruits penetrate the western markets and are found on the supermarket shelves, but many do not.

"An apple a day keeps the doctor away" is an adage that first appeared in print only in 1913, but it is in a long tradition of proverbs and adages relating to food and health. Today health departments worldwide promote the benefits to health of regularly eating fruit and fresh vegetables. The Department of Health in the UK suggests five pieces of fruit each day, to reduce the risk of heart diseases and some kinds of cancer. Particular fruit types rich in anthocyanins are valued especially for their anti-oxidant properties. The high levels of anthocyanins present in *Vaccinium* berries, such as cranberry *V. oxycoccus* and *V. myrtilus*, the bilberry scavenge for damaging superoxide radicals. A relatively new crop, but one of growing significance is aronia (*Photinia melanocarpa*) whose berries have 5-10x the flavonoids and anthocyanins of cranberries, and are also rich in polyphenols, minerals and vitamins. Unfortunately its alternative name, black chokeberry, refers to its astringent taste, but it can be made palatable by mixing its juice with other fruit juices, significantly increasing the potential health benefits of the juice cocktail.

Despite regular admonitions by health departments to eat five or seven pieces of fruit each day for our health we love them and eat them for our pleasure rather than a source of health and energy. In history and iconography many fruits have become particularly associated with a special kind of pleasure, that of love and sex.

Cherry Lips and Figgy Cunts

The seventeenth century equation of an orchard with the Garden of Eden was based on a concept strange to us, that plants were asexual and produced their fruit without earthly sin. In John Parkinson's "*Paradisi in sole Paradisus terrestris*" the only animals shown are also those that inhabit the space between earth and heaven, birds and butterflies.

Ironically the scientific exploration of plant sexuality had begun. In 1671 two great anatomists, Nehemia Grew in England and Marcello Malpighi in Italy, working separately published their accounts. Grew's *The Anatomy of Plants Begun*, first presented in manuscript in April at the Royal Society, was presented again in December, now in printed form, at the same meeting as Malpighi's *Anatome Plantarum Idea* that had been just published in Bologna. By coincidence at the same meeting John Ray the father of English botany was made a Fellow of the Royal Society. Ray's statement that "pollen is the equivalent of the sperm of animals" may have come from Grew, or Bobart, who seems to have proposed it about 1682.

Rudolf Jacob Camerarius the director of Tübingen Botanic Garden made the first detailed observations of pollination in his De sexu plantarum epistola of 1705 and first confirmed this by direct observations. Initially, it was thought that a fluid flowed from the pollen to the ovule to fertilize it. The observations of Camerarius in the early 18th century were very significant because they were the results of experimental manipulations of pollination but they were limited by the quality of microscopes and microscope techniques available. Koelreuter, the son of Tübingen apothecary, inspired by Camerarius, established many of the most important aspects of flower sexuality and pollination. Between 1761 and 1766 he published the results of a series of experimental pollinations, including inter-specific hybridizations, which demonstrated plant sexuality and the process of pollination, firmly establishing the importance of insects in transferring pollen between plants.

Meanwhile Linnaeus was utilizing the sexuality of plants as the basis of his classification system. He believed that the essence of a plant consisted of its fruit, and he developed the sexual system of classification of plants which adopted a ready mechanism by which all known plants could be identified. The Sexual System, the first fruit of which was the publication of *Genera Plantarum* in 1737, relied only on the ability to observe the male and female parts and to observe some very simple features of them such as their number. There were twenty-four classes based on the number of stamens, their relative length and their degree of fusion. Each class was subdivided into orders that differed in the number of pistils. The system became fashionable, not least because it had categories such as *Polygamia* (*Mariti cum uxoribus & innuptis cohabitant in distinctis thalamies* – husbands live with wives and concubines). Naturally there were moral and theological objections raised by some church leaders.

But 17th and 18th century science was only rediscovering an ancient knowledge. True in classical time the role of bees and pollen in pollination was not properly understood: Vergil (70-19 BC), the author of the Aeneid, also wrote the Georgics, an idyllic poem, extolling the rural life, but it also contained a great deal of practical information. The fourth book is about bee-keeping and although he repeats Aristotle's erroneous observation that honey was collected from the honey-dew of aphids, he notes the value of thyme as a honey plant centuries before the role of bees in pollination was established. Nevertheless fruits were linked to reproduction and symbols of sexuality, especially female sexuality.

The type of fruit given by Eve to Adam is not specified though it is frequently depicted as an apple. Alternatively, in some accounts, a pomegranate (*Punica granatum*); the name comes from French (apple with many seeds) is designated. The fruit so full of seeds symbolised the womb and was venerated as part of the cult of Hera the mother of the gods, and goddess of marriage and child birth. Pomegranate juice was a remedy for infertility. Priests of Demeter and Eleusis were crowned with pomegranate branches and in classical Rome newly-wed women wore headdresses made from pomegranate twigs. It was also the fruit of the goddess of love, Aphrodite or Venus. The veneration of pomegranate dates back to Ancient Mesopotamia; an elaborate headdress of a woman buried at Ur is decorated with clusters of golden pomegranates (2500 B.C.E.).

Solomon's crown, and hence all subsequent crowns, was modelled on the fringing calyx at the apex of the fruit. The Chinese also revere the fruit in this way, and it is one of the "Three Blessed Fruits of Buddhism. Pomegranates also figure in the myth of Persephone the Queen of Hades, who was tricked by him to eat six pomegranate seeds before she left the underworld, forcing her to return there for six months each year. Here the pomegranate is linked with the rebirth of the earth each spring. St. John of the Cross made the pomegranate seeds the symbol of divine perfection.

Ezra Pound and Noel Stock mad a poem of a translation of an ancient Egyptian hieroglyphic.

"The Pomegranate speaks:
My leaves are like your teeth
My fruit like your breasts.
I, the most beautiful of fruits,
Am present in all weathers, all seasons
As the lover stays forever with the beloved,"

While the pomegranate represented the womb the fig, *Ficus carica*, because of its shape and texture, represented the external female sexual organs and female sexuality. D.H.Lawrence summarises

"Every fruit has its secret.
The fig is a very secretive fruit.
As you see it standing growing, you feel at once it is symbolic:
And it seems male.
But when you come to know it better, you agree with the Romans, it is female.

The Italians vulgarly say, it stands for the female part; the fig fruit;
The fissure, the yoni,
The wonderful moist conductivity towards the centre.
Involved,
Inturned,
The flowering all inward and womb-fibrilled;
And but one orifice." - D. H. Lawrence

It represents female sexuality in Milton's Paradise Lost and Hebrew texts of the Midrash Aggadah from the fourth and fifth centuries C.E. Both in Milton and the Midrash the profanity of sex and its sacredness as a source of human creativity and imagination are espoused, as if the fruit of the tree of knowledge was a fig that debases Adam but also it is the fig leaves which clothe him after the Fallii as he embraces his sexuality. In a similar way another fig, the sacred bo-tree (*Ficus religiosa*, derives its significance in part from the way as a female element it surrounds and engulfs and unites in marriage the male element the Palmyra palm the kalpa-tree, or the "tree of life" of the Hindu paradise. Many figs are strangling plants. They germinate on the bough of a host, the sticky seed rubbed their by a bird, or left in a bird dropping, and grow as an eipiphyte, meanwhile sending slender roots down the trunk. When their roots reach the soil there is rapid expansion of the fig so that new roots surround, embrace and eventual strangle their host: a very misogynist symbol of female sexuality.

Fico is the Italian for fig and fica is Italian for cunt but for the Greeks like Aristophanes apparently they represent both male and female sexuality. Perhaps this arose from their understanding of fig sexuality. After cultivation and selection for millennia cultivated figs are now dioecious with separate male and female trees. The caprificus or goat fig, produces small flask-shaped syconia, within which there are tiny florets. The syconia of caprificus contain only male florets but small wasps carry pollen from the caprificus to the female florets inside the syconia of the female trees so that the syconia develop and ripen into edible figs. Ancient practice was to hang a bough of the caprificus in the female tree, there is even a verb for it "*caprificare*". Herodotus refers to caprification. Theophrastus describes a similar process by which a male palm inflorescence was placed among the female inflorescences in a date palm to encourage fruit to form Some varieties like the Smyrna fig have only female fruits and require a caprification to produce any fruits. These were imported into California from about 1880 but unfortunately the knowledge of caprification was not transported with them; the figs failed to ripen and fell off until caprification was introduced by Eisen. The first successful crop was in 2000. For every 100 Smyrna figs 3-5 caprifigs are grown to provide pollen. There are other figs that are monoecious and require no intervention.

In a play of Aristophanes play there is the following exchange:

"TRYGAEUS: You shall have a fine house, no cares and the finest of figs. Oh! Hymen! oh! Hymenaeos! Oh! Hymen! oh! Hymenaeos! LEC.E.ER OF THE CHORUS: The bridegroom's fig is great and thick; the bride's very soft and tender."

Hymenaeos, the son of Aphrodite and Dionysus(Bacchus), was the god of the marriage ceremony. Priapus his brother was a fertility god whose huge phallus was generally made of wood of the fig-tree. According to Plutarch, a basket of figs formed one of the chief objects carried in the procession in honor of Dionysus.

The sycamore fig (*Ficus sycamorus*) was revered in Ancient Egypt. Sycamores grew either side of the eastern gate of heaven from which Ra the sun god emerged each morning and was associated with several goddesses. Coffins were made of its wood to signify the return of the dead to the womb. Celebration of the fig as a fertility symbol is not confined to the classical Mediterranean. The bo-tree (*Ficus religiosa*) of the Buddhists is said to derive greater sacredness from its encircling the palm--the Palmyra palm being the kalpa-tree, or the "tree of life" of the Hindu paradise. This connection is termed by the Buddhists "the bo-tree united in marriage with the palm," and we have in it the perfect idea of generative activity, the combination of the male and female elements.

Other fruits have also been suggested as the fruit of the tree of knowledge. Citrus fruits were suggested by some. The characteristic of citrus plants of citrus of bearing flowers and fruit together gave them an Edenic quality. This idea encouraged the spread of that status symbol of the upper classes, special glasshouses, orangeries, in order to keep them in the northern climate. Gerard in his Herball of 1597 reported that Jewish tradition supposed the banana to be the fruit, and Linnaeus also mentioned this. Its obvious phallic shape probably lies behind this though it is not male sexuality that brought the downfall of Adam but sexuality generally and in the shape of Eve, female sexuality. It is ironic that two of the fruits most represented as the fruit of the tree of knowledge are now commonly not sexual at all but produce fruits without pollination i.e. parthenocarpically. This is the case with Kadota figs and most domesticated bananas.

The origin of modern varieties of banana and plantain provides an example of the evolution of seedlessness. There are about 35 species of the genus Musa. The wild ancestors are the species Musa acuminata and M. balbisiana. Wild plants have a pulp that does not develop unless seeds are present but mutant varieties of M. acuminata have a pulp that starts to develops even when the flower is not pollinated i.e. they are parthenocarpic. Further mutation of these parthenocarpic clones gave rise to seed-sterile parthenocarpic clones. This may have occurred independently in *M. maclayi* to give rise to the Fe'i bananas of New Guinea. Further events were the evolution of male-sterility but before this happened hybridization occurred between *M. acuminata* clones and between M. acuminata and *M. balbisiana*. This cross enabled banana to be cultivated in areas of seasonal drought outside the moist tropics of its origin in SE Asia. The hybrids are triploid or tetraploid and are more vigorous.

In comparison to bananas, which have a sweet sugary flesh, plantains are harvested unripe and have a starchy flesh that requires cooking. Cultivation of both bananas and plantains spread though the Old World tropics by the end of the sixteenth century when they were introduced to Central America. By the nineteenth century plantains had entered tropical vegeculture systems to the extent that they had become a staple. For example their easily prepared and digested carbohydrates allowed arrival the population of the Yanomamo to expand.

Some early authors thought that the banana was the fruit of the tree of knowledge. Its phallic shape suggested the fall into sin. Here cultivation of the banana became an expression of American imperialism as in the late 19th century American fruit companies like the United Fruit Company came to dominate the trade. Relatively few clones are grown, exposing bananas to a significant risk of disease from banana wilt (Panama disease), or leaf spot, the pathogenic fungi adapted to attack individual banana genotypes that are grown on a large-scale.

Another fruit crop that also sets fruit parthenocarpically and has a narrow genetic base is the pineapple. Cultivars had already been selected by native Americans before the arrival of the Spanish and Portuguese. Pineapple was used to provision ships And because the crown of the fruit can be used to establish a new plant it was readily established around the world. However many stem from just a few taken initially to France and propagated from there as the fad for growing "pineries" spread.

The practice of propagating fruit varieties by cuttings is not confined to parthenocarpic varieties. In species like apples that are self-incompatible it is the only way to reproduce a favored tree. The seed progeny of crossing between apple trees pippins, are just too variable otherwise. Grafting and cuttings of *Citrus* became popular in the 1800s. Many favored varieties are hybrids that must be propagated in this way. Breeding takes place therefore separately for a good rootstock and for the fruit bearing scion. In *Citrus* self-incompatibility and even parthenocarpy is present in such varieties as the Satsuma and Washington Navel orange. In many others seed is set but by an asexual process called apomixis where a cell surrounding the female organ spontaneously produces a seed; this is virtually essential in varieties that have an uneven number of sets of chromosomes, like the widespread triploids, where a normal sexual reproduction is disrupted.

All this hints at a lack of passion but it is passion and voluptuousness that fruits truly represent. Pomegranates are not the only fruit with a strong sexual image; the shape of many other fruits has been likened to part of the female anatomy. Melons or cocoanuts are comedy breasts. Plum-tree was a catch phrase for the female organ though more frequently today plums refer to the testicles. Cherries have often been linked to feminine beauty. The cherry genus *Prunus* has provided several important fruits cherries (*P. avium, P. cerasus, P. tomentosa*), plums (*P. domestica, P. salicina,* and *P. americana*), apricots (*P. armeniaca*), peaches and nectarines (*P. persica*) as well as almonds (*P. amygdalus*). Different species have been domesticated in China, central Asia, Europe and North America.

Hybridization between closely related species has been, as long with mutation, has been important in the origin of different fruits. Nectarines are simply a hairless mutant of peaches. The poets have been forthcoming about cherry lips ripe for the kissing and cherry fruits ripe for the plucking, love in an Arcadian age. As the Cavalier poet Lovelace put it.

"Love then unstinted, Love did sip
And Cherries pluck'd fresh from the lip
On Cheeks and Roses free he fed;
Lasses like Autumne Plums did drop
And Lads, indifferently did crop
A Flower, and a Maiden-head" - Richard Lovelace

Cherry, is of course also slang for a young girl and virginity. In iconography the cherry is often placed as a counterpoint to the apple, representing the joy and innocence of earthly love before the fall of humanity, against the sorrow of the sinful love after Adam had bitten into the apple of the Tree of Knowledge. Cherry Fairs were markets, perhaps originally held in cherry orchards, where there was boisterous jollity, a festival when a sexual partner could be found.

Herbs and spices and globalization

The arts of the kitchen burgeoned with new cultivated crops. Trade with distant cultures brought a wider variety of produce to utilize. The Romans had available most of the herbs and spices eaten today even in different varieties utilized in various combinations to produce a sauce. Cumin, Ammi (white cumin), Asafoetida, Anise, Basil, Bay leaf, Capers, Caraway, Cardamom, Cassia, Celery seed, Cinnamon, Coriander Seed, Cumin, Dill, Fennel, Ginger, Horse-radish, Juniper berry, Lovage, Mint, Mustard seed, Parsley, White and Black Pepper, Poppy seed, Rosemary, Rue, Saffron, Sesame seed, Thyme and Turmeric. We get a glimpse of Roman culinary arts in the recipes collected under the name of Apicius a fist century gourmand, *De re coquinaria* and *Excerpta*.

The use of herbs and spices to flavor bland foods is an ancient practice. They characterize many different cuisines: the lemon grass (*Cymbopogon citrates*) of Thai food is just one. A large number of herbs and spices come from plants of just three families. Many of the flavors and scents come from the essential oils these plants contain.

The mints are interesting because of the way a few species and their hybrids provide a number of different flavors and smells: Peppermint - *Mentha x piperita* (*M. aquatica* x *M. spicata*), Eau de Cologne mint - *M. x piperita* var *citrata*, Spearmint - *Mentha x spicata* (*M. longifolia* x *M. suaveolens*), Apple mint - *M. villosa*, Pennyroyal - *Mentha pulegium*, Round-leaved mint - *M. suaveolens.*

The essential oils and other compounds that give the smell and flavors are anti-feedants. Some species have been favored in particular as insect repellants, such as Citronella grass, *Cymbopogon nardus*, or natural insecticides, such as Pyrethrum, *Tanacetum cinerariifolium*. Others provide the base scents of many perfumes (Lavender, Patchouli).

What is remarkable is how early many of these spices were traded. The Romans used local herbs and also imported spices: lovage and fennel in sauces; savory in bean dishes; nutmeg and aniseed to preserve meat, and cumin to flavor pastry. It belongs to the umbels (Apiaceae), one of the just three families that provide the majority of herbs. The others are the labiates (Lamiaceae), and composites (Asteraceae). These are three very distinct families and three of the first families to be taxonomically recognized. They are also relatively evolutionarily advanced. Their importance as herbs may be related to their adaptive radiation in the production of insect anti-feedant compounds. The strong tasting compounds discourage insect predators but add savor to our cooking.

The Minoans on Crete used cumin (*Cumin cyminum*) from about the 13th century B.C.E. and there was an extensive trade between India and Egypt. Eudoxos of Cnidus learnt how to sail to India in the 2nd century B.C.E. and returned from their with a boat loaded with spices. Later he led an abortive attempt to reach India by sailing around Africa.

The Romans used three different kinds of cumin from Ethiopia, Syria and Libya, one of these probably *Nigella sativa* the black cumin cultivated in Syria. Many of the spices come from the relative primitive families or come from Southeast Asia or both. Trade in these spices was established to the Near East and China by the Roman era by boat and overland caravan. Pliny's History includes fascinating entries about the source of cinnamon and cardamom. Pliny thought they came from Arabia and Medea, though in fact they came south east Asia and only traveled through these lands on their way to Rome. In later Roman times the trade came to be centered on Constantinople. In the Middle Ages Venice and Genoa controlled the trade with rest of Europe.

| Culinary herbs from the three most important families as sources. ||
Lamiaceae (labiates)	Apiaceae (umbellifers)
Anise hyssop	Angelica
- *Agastache foeniculum*	- *Angelica archangelica*
Basil	Anise
- *Ocimum basilicum*	- *Pimpinella anisum*
Dittany	Caraway
- *Origanum dictamnus*	- *Carum carvi*
Horehound	Coriander
- *Marrubium vulgare*	- *Coriandrum sativum*
Hyssop	Cumin
- *Hyssopus officinalis*	- *Cuminum cyminum*
Mints	Dill
- *Mentha* sp.	- *Anethum graveolens*
Monarda, bergamot	Fennel
- *Monarda didyma*	- *Foeniculum vulgare*
Lavender	Horseradish
- *Lavandula angustifolia*	- *Armoracia rusticana*
Lemon balm	Lovage
- *Melissa officinalis*	- *Ligusticum officinale*
Oregano or marjoram	Parsley
- *Origanum* sp.	- *Petroselinum crispum*
Patchouli	Silphium asafoetida
- *Pogostemon* sp.	- *Ferula* species
Rosemary	Sweet Cicely
- Rosmarinus officinalis	- *Myrrhis odorata*
Sage	**Asteraceae (composites)**
- *Salvia officinalis*	Chamomile
Savory	- *Matricaria recutita*
- *Satureja hortensis, S. montana*	Curry plant
Shiso	- *Helichrysum angustifolium*
- Perilla frutescens	Feverfew
Thyme	- *Tanacetum parthenium*
- *Thymus vulgaris*	Mexican tarragon
	- *Tagetes lucida*
	Tarragon
	- *Artemisia dracunculus* var. *sativa*

They also flavored some dishes with asafetida. Asafoetida is based on three species of *Ferula*, *F foetida*, *F. assafoetida* and *F. narthex*, that grow in the region of Iran and Afghanistan. Its name comes from the Persian word for resin *aza* and the Roman for stinking foetida. It was valued as a tenderizer and preservative of meat. A Persian condiment "the food of the gods" was based on asafoetida and it also has an ancient history in Indian cuisine. In western cuisine it is one ingredient, along with tamarind, that gives Worcestershire sauce its unique flavor. The ex-governor of Bengal Lord Marcus Sandys asked the chemists John Lea and William Perrins of Broad street Worcester to make up the recipe of a sauce for him. The fresh sauce was far too strong but allowed to mature it mellowed into the sauce we know today.

A related product *silphium* or *laser* was from Cyrenaica and was highly regarded by the Greeks, in part because its broad medicinal value, especially as a contraceptive and an abortifacient. It was the gift of Apollo and appeared after heavy rain flooded the coastal plain at about the time the Greeks founded Cyrene in the 8th century BC. It became a major source of revenue for the new colony and was represented on their coinage. Unfortunately it resisted cultivation. By the 1st century C.E. it had been harvested to extinction. Its flavor was so strong that it was stored and transported in flour, Apicius recommended its storage with pine kernels, and these used instead of the resin itself, rather like the way truffles are used to flavor foods today. It was in the course of Alexander's invasion of Asia that asafoetida was recognized as a substitute for silphium. Also called Devil's dung, one way it was not quite as good as silphium was the smell it left on the breath. Several other species of Ferula have medicinal value and the stems of one widespread in the Mediterranean region, called *F. communis*, the giant fennel, which grows 3m tall, can be used as a torch because once lit the pith stays alight. They may have been the original Olympic torches.

The Romans also used pepper (*Piper nigrum* Piperaceae) from southwestern India, cardamom (*Elettaria cardomomum* Zingiberaceae) from India to western south east Asia, cinnamon (*Cinnamomum zeylanicum* Lauraceae) from Sri Lanka, Cassia bark or Chinese cinnamon (from further east), ginger and turmeric (*Zingiber officinale* and *Curcuma longa* both Zingiberaceae) from Malaysia. From the furthest east and the most valuable were cloves (*Eugenia caryophyllus*, Myrtaceae) from the South Moluccas centered on Amboyna, and nutmeg and mace (*Myristica fragrans* Myristicacae) from the Banda Islands just west of New Guinea. Nutmeg is the seed, mace the aril that surrounds it. Originally nutmeg only grew on the 40 sq. miles of the Banda Islands far to the east of the East Indies.

Of these it is peppers, cloves and nutmeg that can justifiably claim to have been two of the most important plants in human history. Pepper was by volume the most significant of these. The Romans were said to be excessively fond of pepper. Augustus sent a special fleet to Egypt and from there another to India to collect spices. The price of spices dropped and pepper became the rage. This southern route was still in use 1500 years later though now centered on Venice not Rome. There was a land based route to Venice via Aleppo that competed with it. From furthest east were cloves and nutmeg were the latest to be imported to Rome but by the 4th century C.E. this was happening. Nutmeg was used as a flavoring, later for example in ale, but also gained a high value as a prophylactic against plague.

The westernization of the world, globalization, began because of attempts to gain and control the eastern spice trade. In the Middle Ages the use of nutmeg spread into Europe. Genoa and Venice supplied northern Europe transshipped via Spain and Portugal by Dutch ships. With the spread of the Ottoman Empire the age-old route of supply from the Spice Islands via Arabia to Egypt and thence to Venice was disrupted. The Portuguese rounding of the Cape of Good Hope and voyages eastward threatened to undercut this Venetian controlled trade, presaging the coming success of the Atlantic based Empires of Portugal, Spain, France, Holland and England.

Now the maritime western European nations, first Portugal and Spain, then Holland and England, sought supplies directly from the Indies and so began their conquest of the world. First the Portuguese and then the Dutch set up a series of way stations and factories to establish the trade. In 1498 Vasco Da Gama traded for spices in India but the further east they could be bought the lower the cost and the higher the potential profit. By 1510 the Portuguese Alfonso de Albuquerque had established a presence at Malacca in Malaysia and pressed on further east reaching Banda in 1512. In 1504 Venetian galleys found no spices in Alexandria or Beirut and by this time Portuguese vessels were carrying spices directly to the Netherlands, Germany and England.

The rediscovery of the Americas by Europeans by Columbus in 1492 was in large part propelled by an attempt to reach the rich source of spices in the Indies by a shorter route. Columbus grossly underestimated the size of the earth. Convinced he had discovered islands lying just to the east of mainland Asia. In his diary he writes of "the very green trees and many ponds and fruits of various kinds". He reported back to his sponsors Queen Isabella and King Ferdinand of the luxuriant tropical forests and the aromatic gum, ginger, cinnamon and pepper he discovered there. But although there were spices they were not the same as those from the East Indies. Half a century later, and by now established on the mainland Gonzalo Pizarro, the youngest half-brother of the conquistador Francisco Pizarro, joined up with Francisco de Orellana in a search for the 'land of cinnamon' in the region of the Upper Coca river in what is now Ecuador. The scale of the expedition was immense, some 200 Spaniards and 4,000 Indians. They found the "cinnamon" trees – probably a different genus *Cinnamodendron*. The constructed a boat to navigate in search for supplies, but in the process while searching Orellana and a small group of 56 Spaniards became separated, or deserted, depending on whose account is believed, and floated down-river, thereby effecting the first passage of what came to be called the Amazon from the hostile "female" tribes people they encountered there. The expedition failed to find a usable supply of cinnamon but Orellana is remembered today in the species name of one very useful plant, the dye plant annatto, *Bixa orellana*.

The Venetian trade was reestablished by the mid-16th century but a pattern had been established First the Spanish and later the English searched for a western route to the Spice Islands. The Spanish had the advantage of a sailor who had traveled to the east with Albuquerque, called Magellan. Although Magellan was killed in the Philippines, his second in command Juan Sebastián del Cano circumnavigated the globe. He returned with only one of the original five ships and a crew of 17 men from the 270 who had started out but the spices the brought back repaid the investment richly.

Over the next two centuries the competition between Portugal, Holland and England for a part of the trade was intense and often brutal. The Dutch East Indies company V.O.C. (Vereenigde Oost-Indische Compagnie) and the East India Company (John Company) vied for command. After the Amboyna massacre in 1629 the Dutch gained unchallenged mastery from Portugal and England in the East Indies and a monopoly of trade there that was to last for a century.

It was finally destroyed when in 1770 a French botanist called Pierre Poivre had smuggled nutmeg and clove seeds out. In 1796 under the pretext of the Napoleonic wars the English invade Banda and transport nutmeg plants to British colonies. Soon an alternative industry was established in the West Indies, especially Grenada. Subsidiary trades were established Asia, initially to finance the spice trade, that in time became more important than the spice trade: cloves for salt in the Persian Gulf, cloves for gold or opium in India, opium for gold, silver and silk and tea in China. Cotton fabrics and china porcelain were carried as well.

Flavors from the Americas: chocolate, vanilla and chili

Flavorings from the Americas include chili peppers (*Capsicum*), vanilla from the orchid *Vanilla planifolia* and cocoa from *Theobroma cacao*. It is a remarkable fact that they were brought together, along with honey, in the drink 'Chocolatl' that was consumed in large quantities as a luxury foaming drink by the Aztecs, in a custom they probably learnt from the south. They believed that the cacao tree was of divine origin. They obtained cacao from Central America by trade. Indeed cacao beans were even a form of currency. Cortes brought cocoa beans back to Spain in 1528 gradually the custom of drinking chocolate spread, especially when chili was left out, sugar, nutmeg and cinnamon added, and the drink served hot. By the mid-17th century the drink, favored in part because of supposed medicinal qualities, had reached England. Both the chocolate makers, Fry's of Bristol and Terry's of York, started as apothecaries. In London around 1700 someone thought to add milk and then in early Victorian times methods to solidify chocolate were perfected.

The Spanish monopoly on the trade in cacao beans was broken by the Dutch when they captured Curaçao and later the French captured Cuba and Haiti. Today Ghana is the major producer though cocoa was first planted there only in 1879. There are three main types of cocoa called Forastero and Crillo and their hybrid Trinitario, which is a hybrid of Forastero and Crillo. Forastero is the most important producing strong flavored beans. The beans are extracted from the pods and cleaned before being allowed to ferment and then dry.

The two other world conquering American flavor plants are vanilla and chili. Astonishingly they share some aspects of the chemistry of their main flavoring compounds which are called vanillin and capsaicin respectively. Vanilla is highly unusual in that it is one of the very few plant products that comes from the largest family of plants, the orchids. *Vanilla planifolia* is from Mexico originally, but is now widely grown in the tropics, in places like Madagascar. Far from having the kind of delicate form we expect from orchids it is a large climbing plant. The flowers are not particularly pretty, but is the pods that provide vanillin, though now much is also produced synthetically. Vanillin is an aromatic compound derived from phenol called 5-hydroxy-3-methoxybenzaldehyde. Natural vanilla contains several hundred other compounds which enrich the taste. Coca-Cola is the largest consumer of natural vanilla. The introduction of "New Coke" in 1985 is one of the great commercial flops. Consumers just didn't like the taste: it was sweeter and synthetic vanillin was used instead of natural vanilla. The economy of Madagascar crashed too. The company soon reverted to the "Real Thing" flavored with natural vanilla and even introduced a Vanilla Coke in 2002.

In the accounts of Garcilaso de la Vega of the expedition of Pizarro and Orellana, mentioned above he notes that the Indians never wish for any other condiment than their uchu, which the Spaniiards call aji, and in Europe pepper. This was the chili pepper *Capsicum baccatum.* There are several other species of pepper and many varieties , varying in their hotness. The heat is conferred by capsaicin a class of alkamides, with a vanillyl group attached to long relatively unbranched chain containing a double bond. The vanillyl group stimulates membrane proteins called VR-1 or vanilloid receptors on cutaneous sensory neurons leading to a massive release of neuro-peptides including those responsible pain transmission, the 'substance P' molecules. One test of pungency is the Scoville test. It consists of diluting an extract of the chili pepper until no heat is sensed. By this test, from data reported by the Chili Pepper Institute of the University of New Mexico, in contrast to Bell peppers (*Capsicum annuum*) that score zero on the Scoville dilution scale, Habanero (*C. chinense*) score 150,000-200,000, Tabasco (*C.*

frutescens) 120,000, Jalapeno (also *C. annuum*) 25,000, Cayenne (*C. annuuum*) 8,000-23,000 and Aji   (*C. baccatum*) 17,000. differences in pungency are not just related to genetic factors but also to the growth conditions. As can be seen in these results there has been extensive selection, especially in *C. annuum* for sweeter cooler types.

Regional cuisines, Mangiare bene or Mei Chi

While Roman cuisine is characterised by its luxury, its use of exotic ingredients and its excess its inheritor modern Italian cuisine has quite a different character. It is characterised by simplicity, by the use of local produce, and by regional variation. Each region, each town has its own specialty. Local produce include ingredients collected from nature like fennel, or asparagus or mushrooms of all sorts. Italian cuisine can be dated back to the first western cookbook, "Art of Cooking" written by Mithaecus, a resident of fifth century B.C.E. Syracusa, then the largest city in the western world. Unfortunately the book is lost but Syracusa was a byword for luxury. Plato

"was in no wise pleased at all with the blissful life, as it is termed, replete as it is with Italian and Syracusan banqueting; for thus one's existence is spent gorging food twice a day and never sleeping alone at night and all the practices which accompany this mode of living."
-    Plato

Italian culinary decadence was is still to the fore in Apicius' works from the first century C.E. The cuisine of Imperial Rome was extremely cosmopolitan, including ingredients imported to Rome from the Empire and far beyond but no doubt there was also a strong local cuisine based on local produce. However, the highly regional cuisine of today must date from the many centuries when Italy was divided into a multitude of small city-states each controlling their own hinterland. In Piedmont *the Fritto misto alla Piemontese* includes meats plus zucchini flowers, mushrooms, cauliflower, semolina, amoretti and apples; in Lombardy the specialty is

Fagiolini (string beans in agrodulce) with *Tortelli di zucca*(Pumpkin); in Veneto a specialty is *Pasta e fagioli* (Borlotti beans, celery, carrot, onion, olive oil, tomato) or *Risotto al radicchio* (Arborio or Carnaroli rice, onion, olive oil, red chicory); in Toscana a specialty is *Infarinata* (beans, onion, carrot, celery, garlic, parsley, corn flour, black cabbage (or curly kale), fennel seeds, potatoes) or *Fagioli all'uccelletto* (white beans, tomatoes, sage, cloves, olive oil); in Lazio there is *Spaghetti all'amatriciana* (spaghetti, tomatoes, chili pepper, pecorino Romano, olive oil) and *Spinaci alla romana* (spinach, garlic, pine nuts, raisins pepper); in Campania there is a Soup *maritata* (broccoli or kohlrabi, turnip greens, savoy cabbage, marjoram, parsley, rosemary, hot red pepper, zucchini (courgette), olive oil, garlic, mint leaves); in Puglia, *Pisella alla pugliese* (peas, pearl onions, parsley, olive oil); in Basilicata, *Lagane con lenticchie* (lentils, garlic, red chili pepper, semolina flour, durum wheat); in Calabria – *Melanzane in insalata* (aubergine, mint leaves, garlic, olive oil, pepper); in Sardegna, *Malloreddus* made from durum flour and saffron; and in Sicilia, *Maccheroni chi vrocculi arriminati* has macaroni, cauliflower, onion, saffron, olive oil, raisins, pinoli, and basil leaves.

The specialties use distinct locally grown cultivars, of rice, beans or brassica, not widely available. For example the broccolis and cauliflowers include: white curd types (Cavolo Broccoli Di Verona Tardivo, from Lombardia, Cavolo Fiore Di Toscana from Toscana and Cavolo Fiore Palla Di Neve from Campania); yellow cauliflowers (Cavolo Fiore Di Macerata from Marche and Cavolo Fiore Verde Di Palermo from Sicilia); green types (Cavolo Broccolo Ramosa Calbrese from Calabria); purple types (Cavolo Fiore Violetto Di Sicilia), pointy romanesco types (Cavolo Broccolo Romanesco from Lazio); nutty flavored cauliflower (Cavolo Fiore Di Jesi from Marche); sprouting broccolis (Sparracedu from Sicilia) and flowering broccoloi (Mugnoli from Puglia). Local land races differ in their color, flavor and the season in which they are grown. In regional dialects each has its own name: broccoli near Lecce in Puglia is 'Munguli' and in Modica in Sicilia it is 'Ciuretti'. Traded only locally only in markets and sown on year by year from locally supplied seed these distinctions have survived into the 21st century though they are under threat from homogeneous white cauliflowers and purple sprouting broccolis grown from seeds supplied by international seed companies for the international trade.

The regional flavor of Italian cuisine is not inward looking. Italy has been exposed to a broad range of cultural influences, from immigrants of all sorts, colonizers or conquerors. I am grateful to Mary Taylor Simeti's wonderful book on Sicilian Food for describing how the various strands and influences from the native Secani and Siculi, Greeks, Phoenicians and Carthaginians, Romans, Arabs, Normans, Spanish have influenced Sicilian cuisine. Sicilian architecture is extraordinary, exuberant baroque combining classical elements in florescent stone, and Norman churches built by Arabs and looking like mosques, and the cuisine is similarly hybrid.

Cuccìa is a pudding made with boiled whole spelt wheat, and Maccu bean soup, that date back to Roman times have already been described. *Pasta con le sarde* is said variously to date back to the Phoenicians or the Arabs. The second is more likely because of pasta was probably their introduction. There are various recipes but the sauce normally combines sardines, raisins, pine nuts, fennel, saffron, parsley and capers.

The period of Arab dominance of Sicily was relatively short but it had a profound influence on its cuisine. Today Sicily is noted for its desserts, a strength that may date to the Arabic influence and their cultivation of sugar-cane. Cassata Sicilina a sweet sponge filled with ricotta cream almond paste and candied fruit, has an Arabic origin. The Arabs also brought rice and citrus fruits. Arancine, rice balls stuffed with chopped meat and peas. The Spanish brought another ingredient for the impanatiglie, the pastries filled with chocolate, dried figs, jam, almonds and even beef, a snack food that could be taken to be eaten on journeys and can now be bought at the buffittieri, hence the word buffet, food available on the street or in little cake and coffee shops that can be eaten on the go or standing up at the bar.

Another plant probably introduced by the Arabs was the aubergine or egg-plant. Pasta alla Norma combines tomato, eggplant and ricotta salata though with these ingredients it does not date back to the Normans, and is instead named after Bellini's opera "Norma" because it so harmoniously combines its ingredients. Rather it was that the rich cuisine of classical times and from the east survived in Sicily through the Norman period, the Norman aristocracy had Arabic cooks, and was introduced from there to northern Europe.

Cuisines became differentiated on class lines and older dishes like Caponata, a vegetable casserole of fried aubergine, with pine nuts, celery, olives and capers in a sweet sour tomato sauce, survived as peasant dishes.

Sicily is not unique, though it stands at the crossroads of the Mediterranean. Claude Levy-Strauss wrote that the most clearly seen traits of a culture are its eating habits. Taking this further it is the range of cooking ingredients, the plants each utilizes that most clearly characterizes it. In every part of the world there is a distinct cookery whether it is national, or regional like the four main regional cuisines of China, Cantonese, Szechuan, Pekinese (northern) and Eastern, or local. In a world that is becoming more and more globalized it is the national and regional cuisines that are surviving best. Almost any city in the world is now a cross-roads and rich with restaurants representing any national cuisine one might want to sample. Even a small local supermarket may today be full of a range of ingredients from around the world that would have astonished even the Romans, but the local cuisines survive.

4 For Strength and Beauty

Throughout human history plants have provided for construction and craft, timber for building and fibers for fabrics and paper. Until very recently and the establishment of petrochemical industries plants have also provided chemicals for a multitude of other purposes. They have provided materials dyes and gums. And they have provided   such as dyes to beautify ourselves and our possessions.

Plant Fibers

The strength of plants comes from cells that have a tough cell wall and lack the living center of a cell, a protoplast, at maturity. These sclerenchyma cells form a dead structural element in the plant and also are part of the xylem tissue, the water conducting tissue, forming a network of pipes through the plant. They have cellulose cell wall in two layers. The inner secondary cell wall is produced inside the primary wall, after the cell has elongated or enlarged. It makes the cell elastic, allowing it to deform, but returning to its original shape after the stress is removed. The cellulose fibrils are regularly arranged, in parallel to each other, mainly longitudinal to the main axis, and in a weave with alternating layers at different angles . The more acute the angle of the fibrils to the main axis of the cell the greater is the stiffness of the cell. Hemp fibers, with a mean fibril angle of 3°, are four times as stiff as cotton hairs with a mean angle of over 30°. However, if the fibrils are orientated more transversely, the cell is less likely to break, because the helically wound fibers can buckle inwards rather than snap when put under strain.

The cellulose network of the secondary cell wall becomes impregnated with a matrix of lignin, a complex polymeric resin which helps to bind the fibrils together, and prevents them shearing apart . Others cell walls are impregnated or encrusted with the fatty substances, cutin or suberin, making them impermeable to water. Some sclerenchyma cells become the receptacles of plant metabolites like resins and gums. Sclerenchyma cells may be found in a specialized sclerenchyma tissue but they are also commonly found interspersed among other kinds of cells in other tissues, as idioblasts. Sclerenchyma cells may be isodiametric (`stone cells' or sclereids) but the significant ones for human use are the elongated tracheids and fibers, which differ in their length and diameter.

Many of the differences in the kind of sclerenchyma cells observed in different species can be related to differences in the efficiency of conduction and mechanical strength. The primitive multipurpose tracheid is imperforate though water can pass through the cellulose of the primary wall in the gaps in the secondary cell wall called pits. Conifers only have these kinds of cells. The angle of the orientation of cellulose fibers changes with age and the strength of conifer wood increases from the inside to the outside of the trunkref1. The outer wood is more recent and produced by a larger plant than the inner. Hence it is likely to be more greatly stressed. The greater strength relates to the greater length of the outer tracheids, to the nearly vertical orientation of the cellulose micro fibrils and to the greater proportion of cellulose. Conifer wood is called by the timber trade softwood but this does not mean it is weak. The presence of a uniform sclerenchyma cell type allows it to be worked rather easily in carpentry.

The wood of flowering plants has several types of specialized cells. The vessel elements have an opening at each end, they are perforate, and are blunt ended cylinders with a direct conduction path through them like the segments of a pipe and so they readily allow the transport of water. Vessel elements tend to be shorter than tracheids. Fibers are narrow and thick-walled and provide strength. The transition from tracheids through fiber-tracheids to narrow diameter libriform fibers is marked by a reduction in the number, size and elaboration of pits. Libriform fibers have small slit like pits oriented parallel to micro fibril orientation, maintaining strength. Although the slit-like pore is too narrow to allow the effective conduction of water, it is enough to allow the maintenance of a protoplasm. Many fibers appear to remain alive for several years. In some cases septa are formed across the fiber after the secondary wall has formed. These septate fibers often contain starch and may have a storage function. The starch may be important as a source of sugars translocated into vessels to increase osmotic pressure and pump prime them, helping to start water flow in spring or after the water column has been broken.

There may be a relationship between vessel diameter and plant height, with taller trees having broader diameter vessels, but the pattern is complicated by the degree of environmental stress the tree is likely to suffer. Trees from temperate and xeric regions generally have vessels with a narrow diameter but this is counterbalanced by a tendency to have a greater frequency of vessels. In cross section the trunks of vines have a lobed or flattened appearance which makes them very flexible. In other plants different sizes of vessel element are often observed in relation to growth rings. A few very wide vessel elements are produced in spring, decreasing either gradually or abruptly in size as the season progresses. This gives rise to the well-known distinction between ring porous timber as in oak, *Quercus*, and diffuse porous timber as in beech, Fagus. However even in so-called diffuse porous species, vessel diameter decreases as growth slows towards the end of the season.

In most cases there is no clear relationship between the height of a plant and the length of individual fibers and xylem vessel elements. Long fibers are more obviously associated with particular taxonomic groups, especially the figworts, Scrophulariales, the nettles, Urticales and the mallows, Malvales, rather than any particular environmental factor or growth form.

Fibers are found in all sorts of plants not just trees and in all parts of the plant, not just the stem. Fibers can be very long, up to 10 cm in hemp (Cannabis), and up to 55 cm in ramie (*Boehmeria*). They gain this length by an extended period of growth after cell division. There are three main stem fiber (bast) sources; flax, jute and kenaf. Fibers from stems and bark differ vary in their chemical composition. Flax and ramie and are mainly cellulose and are white. They provide a fine fiber. Jute and hemp are coarse and brownish, and contain ten to fifteen percent lignin. There are the unicellular hairs of fiber plants, up to 6 cm long in cotton (*Gossypium*). Very hard wearing paper used to make banknotes is made from cotton fibers.

As well as fibers for wood, textile and paper plants provide resins, dyes and pigments with which to decorate them.

Softwoods and Hardwoods

The strength of a timber and many other characteristics of a timber such as stiffness and hardness are correlated to the wood density or specific gravity though not in a straightforward way. The distribution of density across a timber is as important characteristic defining the strength of timber as the overall density. The elasticity, compressibility and resistance to shearing, all important qualities of the timber used for construction, are determined in different ways by the types of cells and their arrangement and not just simply by the density of the wood.

Hardwoods are the wood tissue from flowering plants that are trees and have a mixture of cell types in the wood but always have a high proportion (30-80%) of tightly-packed long thick-walled lignified fibers, along with shorted fatter vessel elements and ± isodiametric living cells. Lignified cell wall material is about 1.5 times as heavy as water but buoyancy is conferred by the air in the spaces in and between cells. There is wide variation in the distribution of cell types, and the evenness of the grain. The qualities of sapwood and heartwood portion of a trunk are different. The heartwood is the oldest part of the trunk in the center. It consists of many dead cells, often packed with substances produced by the sapwood. The sapwood is living and here the water is transported. The heartwood is more resistant to liquid penetration and chemical attack and hence to rot. The durability of timbers varies considerably between species. Softwoods have a lower hemicellulose content and higher lignin content and so are more resistant to chemical attack.

The proportion of the trunk that is sapwood varies between species; it is very wide in *Pinus radiata* but only 1-2cm in eucalypts. In large pieces of timber or sections of veneer the "figure" can be observed made up in part by the grain and also by the color variations in the annual rings and between the sapwood and heartwood; this is obviously influenced by the direction in which the wood is cut, quarter sawn or back sawn. Some timbers have an even grain and others it is conspicuously wavy. Uneven grain timbers when sawn may have a weakness where the cut crosses the grain. It is better therefore to split the wood, rive it, the split follows the grain and produces pieces of wood with greater strength. An even density distributions is also favored for veneer peeling and slicing, though another consideration is the decorative value of the veneer, from the "grain" that is generally conferred by the direction, size and arrangement of cells.

The softwoods are the wood tissue of conifers such as pine (*Pinus*). They are so-called because they have a more homogeneous tissue, a more even "texture", composed mainly (about 90%) of tracheids, cells intermediate between vessel elements and fibers, and consequently they are generally easier to work in carpentry. To produce the flat seat part of Windsor chairs that requires many holes drilled to take spindles without splitting a softwood such as pine is usually preferred. Yellow cypress, white and lodgepole pines and western hemlock are all particularly uniform in density, have a uniform grain, and are therefore excellent for carving and turning on a lathe. In contrast Douglas fir has greater density variation between the late and early wood of each annual ring. The lower hemicellulose and higher lignin content also makes softwoods a better sourced of as a source of wood pulp which can be turned into a variety of materials not just paper.

The specific gravity was important because before of the advent of mechanization it determined how easy wood was to transport. Species with specific gravities much less than 1 float and could be floated down rivers after they were felled. It is a little more complicated than this because wood is hygroscopic, it will absorb moisture in the wet, and also freshly felled timber is "wet" and so densities are normally compared in dry-seasoned or even oven dried timbers. Air-dry densities are taken at a standard moisture content of 12%. After felling the timber shrinks as it dries but there are marked differences between species in the their stability. Walnut (*Juglans regnans*) is exceptionally stable and hence it has been favored for items such as tableware and gunstocks. Various kinds of hardwoods, often called mahogany, were favored for superior furniture and joinery because of their stability. The mahoganies *Swietenia* (S.G. 0.51) from the Americas and *Khaya* from Africa are two of the most used. Mahogany wood is strong, resists rot and termites and has been favored for joinery because it is easily worked. Rosewood *Dalbergia* was another important species.

Differential shrinking can lead to warping of the timber but the consequences of this are different dependent upon where and the direction planks are taken. Those sawn close to the exterior and tangentially to the trunk will warp asymmetrically will those sawn radially retain their shape. Softwoods are not necessarily less dense than hardwoods, some, especially slow-growing ones, are actually harder and heavier than some hardwoods; Pinyon pine (*Pinus edulis*) and juniper (*Juniperus communis*) are examples, they have small densely packed cells. Neither are softwoods generally weaker than hardwoods.

A few species of tree provide dry seasoned timbers that have an exceptionally high specific gravity. Various species of *Diospyros* ebony were favored for fancy veneers and beading; specific gravities vary between 0.960-1.120 kg/m3. The so called ironwoods also sink in water. Several different genera provide ironwoods. Lignum vitae, or the "wood of life" (*Guaicum officinale*) is the hardest of the commercial timbers with a specific gravity of 1.37. Its exceptional density means that it turns smoothly on a lathe. Its density and high resin content make it resistant to friction and abrasion. It is self-lubricating and under certain conditions the wood wears better than iron. Because of this, the wood has been highly valued for pulley sheaves, bearings, casters, food-handling machinery, and for end grain thrust blocks which line the propeller shafts of steamships. The sweet-smelling resin contains 15% vanillin and was highly sought after as a cure-all. Peroxidase enzymes in blood cells oxidize chemicals in the resin resulting in a characteristic blue-green color change. The genus *Tabebuia* the trumpet trees, that are often grown as ornamentals for their glorious trumpets of yellow or pink from S America also provide ironwoods. They are called "quebracho" – axe breakers. The wood is the most durable of any American timber and is reputed to even have been used as the propeller shaft bearings of submarines. *Tabebuia serratifolia* has a specific gravity of 1.20. *Olneya tesota* is desert ironwood (s.g. 1.15). and *Cercocarpus betuloides* (s.g. 1.10) is mountain mahogany. Actually the heaviest plant tissue comes not from wood but from vegetable ivory, the seed of various species of palm.

Amazingly the wood of the tallest living tree *Sequoia sempervirens* is relatively light (s.g. 0.40). This compares with the lightest hardwood American balsa (*Ochroma pyramidale*),with a specific gravity of only 0.17. It is lighter than the bark of Cork Oak

Quercus suber (s.g. 0.24). The main area of cultivation is in the western Mediterranean, especially Portugal and Spain. The outer bark is stripped by hand for the first time when the tree is about 20 years old and then about every 10 years. After allowing it to weather in the open air for six months corks are cut. The waste is turned into all manner of materials such as cork tiles.

Some species have been particularly favored because of their rot resistant qualities. Foremost of these is Teak (*Tectona grandis*). Teak is native to India and South East Asia. They had been used for centuries to construct Arab dhows and Chinese junks before discovery by Portuguese, Dutch and English sailors repairing their ships though it is very heavy and unseasoned logs sink its dry seasoned specific gravity is 0.630-0.720 kg/m3. It was used for the keel of HMS Victory. Strong, durable, resistant to acid and alkali and resistant to insect attack and fungal decay it was highly prized. It has a natural oil but is so hard ordinary nails cannot be driven into it. Trees up to 50m tall and 12m in diameter used to be felled but natural forests have been extensively exploited and trees grown on plantations that have been established in India, the Philippines and Indonesia are felled before they are of any great age. When chosen for felling the trees are at first ring-barked and allowed to dry out as they die in situ, making the logs light enough, after cross-cutting to be dragged by elephants and floated down rivers to the sea-ports.

Resistance to marine borers was maritime construction. Another one was Satinay (*Syncarpia hillii*) from NE Australia. It was utilized for dock piles both in the Suez Canal and in the repair of the London Docks after the Second World War but now is largely protected as an important component of the Fraser Island world heritage site. Similarly greenheart (*Ocotea rodiaei*) a relative of camphorwood was used in the construction of the Panama Canal because of its resistance to termites and other borers. Karri (*Eucalyptus diversicolor*), spotted gum (*E. maculata*), blackbutt (*E. pilularis*), and jarrah (*E. marginata*) are favored for furniture and wood carving. Jarrah is stronger and more durable than oak and resistant to termites and marine borers and has a reputation for use as marine piles and planking. Karri can grow to 100m.

Timber!

It is impossible to over-estimate the importance in human history of wood and timber as a constructional and craft material. The Hebrew name for tree is the same as timber, erez. The great size and strength of the cedar is described many times in the bible. It has been venerated but also a valued timber. Even Gilgamesh after triumphing over his enemy Humbaba he cuts down the forest to make a great cedar gate for the city of Uruk. King Solomon used cedar to build the temple at Jerusalem, buying timber from King Hiram of Tyre and floating them down the coast. The name comes for the Greek *kedros* for juniper or more generally sweet-smelling resinous woods. The resin confers greater durability. Sharing this resinous smell and hence called the Western Red Cedar *Thuja plicata* grows in the Rocky Mountains of Canada and the USA. A magnificent tree growing up to 70m in height with a tall straight trunk its timber was very valued because of its freedom from knots and its strength despite its light weight (with a specific gravity of 0.38). It had multiple uses for the native Americans such as the Haida tribe; timber for construction, for war canoes, totem-poles, the bark for fiber in everything from ropes to clothes the twigs for arrows, the occasional knots for fish hooks.

Wood and timber was the main constructional and craft material before the industrial revolution. This multiple use cannot be over-emphasized in cultures without alternatives such as steel or plastics. A sample of small items, like bowls and bows, were lost and preserved beside the Somerset Trackways, themselves constructed from branches across the soft ground, from the English Neolithic emphasize the versatility of wood. The most important value of timber was for the construction of the vessels that established trade links across vast distances. Phoenician traders opened up the Mediterranean cultures to each other. Arab dhows linked Africa and the Middle East with Asia. The craft that permitted first colonization and then trade across the vast distances of Polynesia were constructed of many timbers but at their heart were vesi logs (*Intsia bijuga*) from Fiji, hollowed because of their great density. Wooden ships made the modern world.

The wide range of uses construction, craft and as fuel has threatened the survival of favored species. This is not just a recent obsession. Deforestation, also to clear land for cultivation, brought environmental degradation. There is evidence of rapid soil erosion in Jordan from 6000 B.C.E. . Concerns about the loss of the cedar forests appear in the guise of Enlil the chief Sumerian god appointing Humbaba to protect the cedar forest from the needs of human civilization. Soon timber was so scarce that it had to be imported into Mesopotamia from western India and the Mediterranean. The cedars of Lebanon were especially prized and vulnerable – timber could be floated down to the Mesopotamian cities. In 450 B.C.E. Ataxerxes attempted to preserve the surviving cedars. The loss of woodlands was not just an economic concern; timber was used in large quantities to build the navies that were increasingly required for trade and for colonial and imperial expansion and security. Egypt was especially reliant on timber from outside its own territory and perhaps it was this, its relative lack of a navy that limited its imperial expansion.

Not so with the Phoenician, Cretan, Greek, Carthaginian and Roman civilizations which each in turn created navies to spread their dominion across the Mediterranean basin. The Peloponnesian War (432-412 B.C.E.) intensified deforestation of mainland Greece. By the time of the Roman Empire much of the Mediterranean basin was deforested. Surviving forests like those on the slopes of Etna came under increasing pressure. With the rise of the Arab empire timber was sought on Sicily and other Mediterranean islands. The demand was huge. For the attack on Constantinople in 717 C.E. the Muslim caliphate collected together a fleet of 1800 ships. Later the Venetian empire required large quantities of timber for the Arsenale depleting the timber supplies of the Dalmatian coast and trying to prevent woodland clearance in colonies like Crete . The centers of power moved west to where there were still substantial woodland in the Iberian peninsula.

In northern Europe extensive tracts of woodland survived longer. Julius Caesar reports finding extensive forests in Gaul. In Britain it is traditional to consider the Anglo-Saxons as the main clearers of the virgin woodland, the wild wood, in Britain, but as far as they did clear any woodland they only continued in a long tradition. A continuous supply of wood of different sorts became essential. The woods were coppiced to provide a continuous supply of wood for Roman industry. Especially important were the iron workings of the

Weald and the Forest of Dean. The extent of the Roman workings in the Weald suggest about 15% of the Weald would have supplied their needs. Only 23,000 acres of coppice wood would have provided a similar need. The distinction between timber and wood is recognized in the Latin words *meremium* (= timber) and *boscus* (= wood) to describe it. One species which coppices well is the *Castanea sativa* (Chestnut) . The Romans probably introduced it but perhaps not for coppice but for its nuts. *Carpinus betulus* (Hornbeam) also began to be an important tree during the Roman period. Roman towns had an insatiable requirement for fuel. Hornbeam was one of the best woods for this purpose. A trade in fuel wood was established. Coppices of Hornbeam were still supplying London with much of its fuel in Tudor times.

It was mostly the secondary woodland and scrub which they cleared. They did not have to contend with the huge trees of the wildwood. The surviving areas of virgin woodland continued to change as they were utilized as a source of fuel-wood, timber or for grazing. King Alfred recorded

"We wonder not that men should work timber-felling and in carrying and building, for a man hopes that if he has built a cottage on laenland of his lord, with his lord's help, he may be allowed to lie there awhile, and hunt and fish and fowl and occupy laenland as he likes, until through his lord's grace he may perhaps obtain some day boc-land and permanent inheritance" .

The use of woodland is advised by King Alfred,

"And I gathered for myself staves and props and bars, and handles for all the tools I knew how to use, and crossbars and beams for all the structures which I knew how to build, the fairest pieces of timber, as many as I could carry..nor did it suit me to bring home all the wood, even if I could have carried it. In each tree I saw something that I required a home. For I advise each of those who is strong and has many wagons, to plan to go .... and load his wagon with fair rods, so he can plait many a fine wall, and put up many a peerless building, and build a fair enclosure with them."

It was the pressure of grazing on common land that brought

about as much deforestation as assarting or felling for timber. In the Domesday Book of 1086 C.E. only 3.4% of the land is recorded as woodland, almost entirely as either coppice woodland or as wood pasture. By 1300 much of central England was an open sea of arable fields and there was intense pressure on surviving woodland. The wood boundary was marked by a wood-bank and ditch and protected by law. A mixture of fenced coppices and open plains with pollarded trees which are open to grazing by cattle was created in some places. Coppices were intensively managed to provide under-wood for fuel, poles, rods and withies. Amongst the coppice some standard trees were maintained to provide larger timber. Coppicing was carried out on a short five to seven year cycle in the Middle Ages. Longer coppicing cycles of 10-17 years became more common from the sixteenth century onwards and fenced to keep out deer and cattle for the first half of the coppicing cycle. The use of woods as particular kind of pasture especially for pigs but also for deer and cattle was well established. Pannage, the right to fatten pigs in the woodland was one way in which the extent of the woodland was measured, as a wood for so many swine. In pasture woods the trees were pollarded so that the tender branches which provided poles and rods were protected from grazing by growing on a short trunk above grazing height and cut every 10-20 years. Gradually pasture woods had a tendency to be converted to open grassland as woodland regeneration was prevented. This is a familiar pattern throughout the world.

The requirement for timber could be filled, generally from sustainable cropping, most trees were felled after 25 to 75 years, but. even a medium-sized timber framed farm house used hundreds of trees. One from Suffolk built in 1500 C.E. used 330 trees. Half of these were less than 9 inches in diameter and only 3 exceeded 18 inches in diameter. Trees older than 75 years were difficult to fell. transport and cut up into usable timber but were required for major projects like the construction of the extraordinary Cathedrals. It became necessary to obtain timber from standard trees left to grow amongst the coppice or in fields and along field margins. The wildwood could no longer provide them. Trees of great age had disappeared from England. When a great construction project was taking place, like the construction of a cathedral, the countryside was scoured for suitable trees. The octagon of Ely cathedral required

16 struts 40 ft. long and 13 inches square but only 10 could be found. The design had to be modified to include some shorter ones. Large timbers were also required in windmills for the post and prices were high, probably several thousand pounds in today's prices, because of scarcity. Even where the construction was stone or brick-based roof beams were required and in the building process itself huge quantities of timber were utilized to build scaffolding. Fencing had to be maintained to prevent the development of wood pasture. In 1483 the first statutes were introduced in England to protect woodland so that woodland had to contain a minimum number of standard trees.

There is no better example of the value of timber in construction than the tremendous wooden ships that have been built at many ages. In the British navy oak *Quercus* was the favored species. Pine was also important. The maintenance of timber supplies to supply the British navy shaped aspects of foreign policy throughout the 18th and early 19th century; keeping the Baltic open for trade in the Napoleonic era, gaining Minorca in the Mediterranean and so forth. The timber industry was a major element in the opening up of new lands to western commercialism. The major difficulty was transporting the logs. Large teams of oxen pulled logs into place. Logs were cleaned of branches and tapered at one end before sliding down to rivers on tracks of branches laid sideways. Other techniques included damming of streams to create a lake holding the felled logs. Breaching the damn made a flood that carried logs down, to a river. This was a technique well known to the Romans. It was still was still widespread in the 19th and early 20th century; about 3,000 dams for the transport of Kauri were constructed in New Zealand. The largest felled trees were cut into manageable planks in situ by the construction of a sawpit, the sawyer on top and the pitman below, his eyes and ears, nose and mouth constantly filling with sawdust.

In each place the larger and most favored timber species were the first to be logged. The exploitation of the timber of Americas east coast was a major motivation for British colonization. As the loggable forests of New York and Massachusetts were exhausted, attention turned first to the saint John River System and the saint Lawrence River and Ottawa River system. White pine was a favored timber because it floated well, while oak, which had a premium value had to be supported on these rafts. Timber was the primary export with, for example 1200 ships carrying timber away from Quebec in one year alone. O the return leg from Britain the ships carried goods and people to colonize the newly cleared lands. Reliance on this export trade encouraged Canadian loyalty to Britain at the time of the American revolution. In the USA attention turned at first to the Mid-West and then the west coast and northwest. Ponderosa pine, western larch, Idaho white pine from the upper mid-west. In British Columbia Western hemlock, Western red cedar, Amabilis fir, Sitka spruce, Yellow cedar and Cottonwood.

In Australia the Red Cedar *Toona australis* were particularly favored. It was likened to mahogany. It had a pleasing deep red color and was durable and easily worked. In addition it floated and so could be easily transported by river. In addition it was deciduous and so could be spotted easily among the evergreen eucalypts. The search for exploitable resources of red cedar determined in large part the pattern of European settlement in eastern Australia One of the saddest tales is that of the penal colony on Norfolk Island, founded to provide a supply of the Norfolk Island pine *Araucaria heterophylla*, even though this timber was not particularly useful. Its establishment was spurred by the lack of suitable timber species around Sydney; they were mostly knotty and resiniferous eucalypts that were difficult to work. But not all penal colonies were so unsuccessful. The Macquarie Harbor colony was established in 1815 to exploit Huon pine, *Lagarostrobus franklinii*.

The Kauri, *Agathis australis*, from Australia, but especially from northern New Zealand was favored for masts and spars and because of its clean even grain and in New Zealand 1,200,000 hectares of Kauri forest were reduced in time to just 4,000 hectares. Logged land could sometimes also be sold for agricultural use. However in many areas the soils proved to be too poor and secondary forests became established. It has been estimated that America had lost half its standing volume of tree timber by the end of the 19th century . Many tree species were lost altogether from the USA . Mechanization, the development of the chainsaw, the skidder, caterpillar tractor, and eight wheeled log wagon enabled year-long logging and brought costs down. The skidder was a winch with long cables that were attached to logs to drag them to where they could be loaded onto wagons but in the process the dragged log would flatten anything , including sapling trees , in its path.

Exploitation of premium timber species moved around the world and was established at different times. Tropical timber exploitation dates as far back as the trade in timber from western India to the Mesopotamian civilizations. Timber species included teak, (*Tectona grandis*) and satinwood (*Chloroxylon swietenia*). Western discovery of the Americas opened up a whole new world of tropical timbers. The most important by far was mahogany (*Swietenia macrophylla*) that could be used in diverse ways from ship-building to delicate Chippendale and Hepplewhite furniture. Cortes used it for shipbuilding and by 1584 it was being used to decorate the Escorial Palace in Madrid . Its advantages are its relative lightness, stability and workability. As the large available trees were cut out on the Caribbean Islands attention turned to the mainland, hence it is sometimes called Honduras mahogany, and then to South America.

Commercial exploitation of West African timber did not really get underway commercially until the 1880s, encouraged by the increasing scarcity of American Mahogany, and in the wake of the exploitation of wild harvested products of copal and rubber but large-scale exploitation of interior rainforests did not take place until the 1950s. Exploitation was almost entirely focused on African mahogany (*Khaya ivorensis*) and to a lesser extent sapele (*Entandrophragma species*). Khaya has very similar characteristics to *Swietania*, though it is a little less stable. Sapele is popular for veneers because of its striped appearance due to a grain that changed direction at irregular intervals. Unfortunately this feature also made the wood liable to warp and awkward to work. Another premium African species was Afrormosia (*Pericopsis elata*) a strong and hard wood that was favored for flooring and ships decks. Unlike khaya and sapele it is a savannah tree.

Most of these timbers have been, and are being exploited relentlessly from the wild. Some of the most favored have become very rare as a result. *Afrormosia*, which was more expensive than khaya was made nearly extinct in West Africa and stocks of sapele are largely exhausted. Exploitation of tropical hardwoods has in many areas been highly selective with only large premium tree species are felled. The intensity of exploitation varies very greatly depending upon the frequency of selected trees of adequate size. It is for example three time greater in South East Asia where denser stands of dipterocarp species are present, than it is in West Africa, where khaya and sapele are dispersed and relatively infrequent. In the latter exploitation amounts only to about one large tree felled per hectare but unfortunately this does not indicate the extent of damage to the forest that logging causes. Surrounding trees and saplings are destroyed in the process of felling and dragging trees to the road. Most importantly roads established to truck timber out provide a route in for pastoralists and agriculturalists to follow, and remaining areas of forests are soon cleared and burned to enable cultivation of crops or for the establishment of grasslands. Different kinds of sustainable management systems have been attempted but with varying success. Selection systems actually destroy non-commercial species to limit the competition to young commercial trees so that they achieve harvestable size more rapidly. The Tropical Shelter Wood System (TSS) clears lines and undergrowth and middle size uncommercial species. Enrichment planting is the planting of saplings of commercial species raised in nurseries elsewhere, though commonly the introduced saplings are soon smothered by weeds and woody climbers.

With the loss of primary forests around the globe there has been a turning to the sustainable management of the forest resource. Management of timber sources is a very ancient practice. Some of the best evidence from the Neolithic of England comes from the wooden trackways laid down across the marshes of the Somerset Levels connecting surrounding higher ground with the low islands of Edington Burtle and Westhay Island. The earliest, the Sweet Track, is the most sophisticated. It dates from about 6000 years ago. It was raised walkway of *Quercus* (Oak), *Tilia* (Lime) and *Fraxinus* (Ash) planks was held up on a trestle of *Fraxinus* (Ash) and *Corylus* (Hazel) pins, all held in place by a foundation rail of Elm. The ash and hazel pins are 1-1.5 m long and straight with a diameter of about 40-50 mm. They were evidently produced from coppice stools which were harvested on a seven-year cycle. The entire trackway, more than 2 km long, was probably constructed in a single year with all the trees felled at the same time. A sample of small items, like bowls and bows, were lost and preserved beside the Somerset trackways.

Modern forestry around the world has been shaped by the activity of individual collectors and landowners, introducing sometimes successfully, non-native trees to a region.

For example in Britain it can perhaps be dated back to the introduction of the *Larix decidua* (European Larch) in the early 1600s from the Alps. Large plantations were established on the estates of the Duke Of Atholl from about 1800. By 1832 there were 10,000 acres of 14 million trees. Not adapted to Britain's mild climate it suffered from larch canker or die-back, though a variety from Poland and Czechoslovakia performs better but was not planted after 1937. Instead *L. kaempferi* (Japanese Larch) or their hybrid became favored. Larch is still favored in polluted areas because its deciduousness confers some resistance. *Picea abies* (Norway Spruce) proved useful in frost hollows on wetter patches of the plateau lands but here the root system is shallow and so it is liable to wind-throw if the site was exposed. *Picea sitchensis* (Sitka Spruce), first introduced by David Douglas from North West America in 1831, has become the main tree of blanket peat throughout the British Isles. By 1965 600,000 acres had been planted and planting was continuing at a rate of 30 million per year. The *Pinus contorta* (Lodgepole Pine) has proved the best pioneer species. It withstands exposure well and is preferred over spruce in frost hollows. *Picea sitchensis* (Sitka Spruce), *Tsuga heterophylla* (Western Hemlock) and *Pseudotsuga menziesii* (Douglas Fir) are liable to be checked in their growth by the early competition of *Calluna vulgaris* (Heather).

The management of tropical hardwoods is still in its infancy. The premium value of teak has resulted in its over-exploitation throughout South East Asia and to the establishment of Teak plantations. Teak is native to Myanamar (Burma) from where it is still exported and Thailand, where it is protected, as well as India and Sri Lanka. Now there are numerous plantations not just within its native range but also in Africa and Central America, many advertising for western investors.

Technological innovation has vastly broadened the range of timber species that are now valuable commercially. Plywood, chipboard and fiberboards of all sorts combine wood fibers and veneers of different sources and overcome the lack of stability of some timbers. The Ancient Egyptians made a kind of plywood by sticking wood veneers together. Chipboard was first developed about 50 years ago; it can combine with resin fibers from flax and sugar cane with wood shavings mainly from softwood species such as Norway spruce (*Picea abies*). Fiberboard is produced from ground wood pulp, from a mixture of hardwood and softwoods, treated with heat and pressure. *Eucalyptus* species. like *E. saligna* from Brazil, are an important source of the pulp. Some regrow after they are cut back and hence provide a repeatedly available source of wood fibers. MDF (medium density fiberboard) has become ubiquitous because, lacking a grain, it can be sawn and drilled and machined without danger of it splitting.

In the tropics and warm temperate regions such as the Mediterranean basin the genus Eucalyptus has been favored to provide shelter belts as well as fast-growing timber. In many areas they have become naturalized and pose a threat to native vegetation. The red gum *Eucalyptus camaldulensis* is a prolific seeder and initially lacking any native pests or diseases became a weed in California until in the late 1990s an Australian defoliating psyllid pest called *Glycaspis brimblecombei* became established. There are about 450 species of eucalypt native to Australia. Their wood of different species varies from relatively soft and light to hard and dense.

Sacred Groves

Trees of particular species have been revered for their size and strength. The tallest tree Sequoia sempervirens "stratosphere giant" was only discovered in the 1990s and is over 112m tall. At least 26 trees over 110m have been discovered, most in Humboldt State Park. At one time even these were exceeded by the Douglas Fir *Pseudotsuga menziesii*, one of which was measured at 117m but now none greater than 99m feet is known. *Eucalyptus regnans* was probably the tallest flowering plant from Southeastern Australia. A stump called the Bulga stump rivaled the girth of the giant sequoia. With the renewed appreciation of "wild" landscapes, trees have come to symbolize nature and forests and woodlands re-established as the home of the gods and as places of religious veneration. Some trees had an especial significance. Part of the Epic of Gilgamesh which is the oldest written story dating from somewhere between 2750 and 2500 BC takes place in the Cedar Forest. Gilgamesh quakes with fear as he enters the forest.

"They stood at the forest's edge, gazing at the top of the Cedar Tree, gazing at the entrance to the forest....Then they saw the Cedar Mountain, the dwelling of the Gods, the throne dais of Imini.
Across the face of the mountain the Cedar brought forth luxurious foliage., Its shade was good, extremely pleasant."

The veneration of the cedar groves continued into the Christian era; Maronite Christians held an annual festival amongst them. The cedar has a special place in the Near East. The origin of the word Cedar is probably from a Semitic root word meaning power but in the bible in Ezekiel 31 God likens Pharaoh's power and pride to that of the cedar but tells him he will bring him low.

"Behold, I will liken you to a cedar in Lebanon, with fair branches and forest shade, and of great height, its top among the clouds. The waters nourished it, the deep made it grow tall, making its rivers flow round the place of its planting, sending forth its streams to all the trees of the forest. So it towered high above all the trees of the forest; its boughs grew large and its branches long, from abundant water in its shoots. All the birds of the air made their nests in its boughs; under its branches all the beasts of the field brought forth their young; and under its shadow dwelt all great nations. It was beautiful in its greatness, in the length of its branches; for its roots went down to abundant waters. The cedars in the garden of God could not rival it, nor the fir trees equal its boughs; the plane trees were as nothing compared with its branches; no tree in the garden of God was like it in beauty. I made it beautiful in the mass of its branches, and all the trees of Eden envied it, that were in the garden of God."

Sacred groves are areas of protected forest. They are widespread around the world, in Ethiopia, India and Europe. The bible is full of references to sacred groves. Abraham plants a grove in Beersheba to call the name of the Lord but later to establish the true religion one of the things Moses is told to do is to cut down the sacred groves. This imprecation is repeated several times in Deuteronomy. In Judges it is related how the children of Israel *"forgat the Lord their*

God, and served Baalim and the groves" (Judges 3:7). Baal was a local god, the genius of place, often represented as a god of fertility, associated later with the rain god that ended the summer drought. But the significance of the Hebrew execration of sacred trees was because they represented an alternative channel of communication with the heavens: rooted in the earth their branches reach to the heavens.

This symbolic importance is widespread in human cultures. In Central America, especially for the Maya, the Kapok tree, *Ceiba pentandra* is the tree of life, Yaxche. Its roots reach to the underworld and the branches hold up the heavens. It flowers in the evening and a night time. It is exceptionally tall, with a long unbranched trunk that, pierces the forest canopy, reaching up to massive branches that spread in the ritual cardinal directions. In the mountains of Chiapas the *Ceiba* is joined by the pine that clothes the slopes. Ritual spaces are decorated by boughs cut from trees and the ground covered with pine needles.

In New Zealand surviving individual trees have been given names. Te Matua Ngahere and Tane Mahuta, Father and Lord of the Forest respectively, are two of the largest Agathis australis or kauri trees that have survived logging. Tane Mahuta is 51 m tall and has a girth of 14 m. Te Matua Ngahere is broader in a girth. One measured in 1856 but no longer surviving had a girth of 23.43 m. These trees are ancient. Tane Mahuta is estimated to be 1200 years old. They reach up above the canopy of surviving trees. A Maori legend tells how Tane-Mahuta placed his shoulders against his father Ranginui the sky god and his feet against his mother Papatuanuka the earth god and pushed them apart making a space in which human beings could live.

The Date Palm (*Phoenix dactylifera*) was venerated in Ancient Egypt as the symbol of eternity. It has identified as the Tree of Life from the Garden of Eden. For Christians too it is the symbol of triumph over death, which is the significance of Palm Sunday and the entry of Jesus into Jerusalem. In other places different tree species represent the tree of life. In Scandinavia it is Yggrasil or Ash tree is represented as the axis of the world. The middle world, the world of humans is held in the lower branches. The Yew *Taxus baccata* is widely regarded as the tree of life because of its longevity and was so regarded by the Celts and Romans and was also revered by native Americans.

Sacred groves of trees were often associated with religious sites. In classical times the temple of Apollo at Delphi was surrounded by a sacred grove of laurel trees (*Laurus nobilis*). Daphne daughter of the river-god was chased by Apollo but changed into a laurel by her mother Gaia to protect her. The priestesses Pythis chewed the leaves to see divine visions though since these are only the bay leaves we use to flavor our cooking she was not getting a "high".

There were sacred groves of Aesculapius at Epidaurus and of Argus in Laconia1. There was a variety of species: and a sacred grove of plane-trees at Lerna and another sacred to Zeus, known as the Altis, at Olympia, at Colophon in Ionia was a grove of ash-trees sacred to Apollo, and a sacred grove at Lycosura included an olive-tree and an evergreen oak growing from the same root. The Grove of Dodona, of the evergreen oak *Quercus ilex*, at the foot of Mount Tomarus was dedicated to Zeus; priestesses called doves, Pleiades, cooing like birds presented oracles by interpreting the rustling of the leaves in the wind. Jason's Argo was constructed from oak timber from Dodona and hence had the gift of prophesy. In 1640 James Howell published an allegory called "*Dodona's grove or the vocall forrest*", rich in tree lore. Christians cut down the last surviving holy tree in 391. Lucky for them they were not attacked by forest nymphs called Dryads, who wore wreaths of oak leaves and danced among the sacred trees, attacked anyone that damaged the trees with an axe. The Hamadryads were nymphs integral to the trees and died when the trees were felled.

In India sacred groves are areas of forest protected by the indigenous population as the home of gods and there are voluptuous tree nymphs too in India called Vrikshaka. The Bo tree *Ficus religiosa* or Pipal, Peepul or Pipul tree has a special place in Buddhism. In the 6th century BC in northern India Prince Gautama nearing the stage of enlightenment sat down under the tree vowing not to rise until he had achieved enlightenment or Bodhi. He was tempted by the evil one Mara and when this failed assaulted him with flaming rocks, but these turned into petals around Prince Gautama. As his meditation deepened the tree rained red petals and at dawn Gautama was transmogrified into the Buddha. In the second century BC a princess who converted to Buddhism took a cutting to

Anuradhapura in Sri Lanka. A descendent of that cutting survives today. In nature peepul starts life as an epiphyte, supported by a host and sends roots down to the ground eventually strangling its host. It has heart shaped leaves with a long leaf tip. This kind of tip is common in tropical trees but not in figs: it is a drip tip and helps the surface of the leaf drain and keep dry so that algae, mosses and liverworts cannot grow on it. It produces purple figs. The leaves, bark and roots are used for a variety of ailments. The Bo tree is also sacred to the Hindus and was one of the earliest trees to be depicted in Indian art and literature. Another sacred tree in India is *Mesua ferrea* (ironwood).

The Grove of Aricia near Rome was associated with Diana. The ancient custom of plucking the Golden Bough from a tree in the Grove of Aricia was taken by James George Frazer as the title for his seminal work of anthropology The Golden Bough . The golden bough represents a message from the heavens. Frazer follows the suggestion of Pliny the Elder that the golden bough was mistletoe. In his description of the Druids he relates

"Anything growing on those trees [oaks] they regard as sent from heaven and a sign that this tree has been chosen by the gods themselves. Mistletoe is, however, very rarely found, and when found, it is gathered with great ceremony and especially on the sixth day of the moon. They prepare a ritual sacrifice and feast under the tree, and lead up two white bulls whose horns are bound for the first time on this occasion. A priest attired in a white vestment ascends the tree and with a golden pruning hook cuts the mistletoe which is caught in a white cloth. Then next they sacrifice the victims praying that the gods will make their gifts propitious to those to whom they have given it. They believe that if given in drink the mistletoe will give fecundity to any barren animal, and that it is predominant against all poisons."

Apart from this we know practically nothing about the Druids except that they were the Celtic priestly caste. Druids were said to occupy the deep recesses of sacred groves. The word Druid is the Gaelic for magician and was thought by Pliny the elder to have the same root as the word for oak *dru* and *wid* knowledge. The ogham script of carved straight lines in groups used in fifth century Ireland and hence Celtic has twenty main letters each of which is also the name of a sacred tree. The oak tree was sacred to the Celts and mistletoe has the right characteristics to have a heavenly role: it is not rooted in the earth but has a holdfast that penetrates the vascular tissue of the host. It grows golden after being plucked. Mistletoe is strongly associated in myth with thunder, the sound of the gods. *Viscum album* is parasitic on oaks but also apples and poplars. The berries are made in bird-lime used to trap small birds. Tropical and New World mistletoes belong to a different family the Loranthaceae do not have these associations though some species have the remarkable ability to mimic the plant which they are parasitizing. Different kinds of parasite arise from the roots of other plants and are associated with the underworld. One called *Striga* is the witchweed because it parasitizes the roots of other plants.

In orchards and gardens the Sumerians and Babylonians sought to recreate a piece of sacred nature. From these Persians the constructed "*Paradeios*", meaning "beside god", gardens with a central fountain and channels flowing along each of the cardinal points, representing the four rivers flowing out of Eden. Temples, from the earliest times provided a sacred grove; the forest of pillars is a stone model of a sacred landscape . The columns represented the trunks of trees and the capitals their leaves. Vitruvius writes that Greek and Roman temple columns were first made of tree trunks and only later of stone. Columns taper like tree trunks and are fluted like bark . The very early Doric Temple C at Selinunte in Sicily constructed between 580-550 B.C.E. had stone outer columns and wooden inner ones. The architect proudly boasts of the stone construction used in Sicily for the first time. In Egypt the palm-leaf form of capital was used for millennia. Capitals are carved with the leaves of other plants too.

This shift from wood to stone sacred spaces is mirrored in the history of other sacred spaces. Ritual Mayan sites have a forest of tree stones and on top of some of the Mayan pyramids there are slots for the trunks of great felled trees.

The henges of northwestern Europe date from 3,500-3,300 B.C.E. considerably earlier than the temples of classical Greece. At a time when the surrounding landscape of Salisbury Plain was becoming cleared of surviving fragments of woodland, artificial groves, sacred circles of wooden columns, tree trunks erected in pits, perhaps partly roofed, were being constructed. Such wooden sacred circles as Durrington Walls and Woodhenge were precursors of the stone circles of that culminated in the magnificent monuments of Avebury and Stonehenge. In the same way the columns of cathedrals are the tree trunks of a sacred grove. In Gothic cathedrals they soar and reach across, to roof the sacred space. It is a familiar idea how cathedral-like the interior of an ancient beech wood is, but this is to get it the wrong way round; it is how like a beech wood the interior of a cathedral is.

Fibers

Fibers come from the stem and bark, from leaves or from seeds. A large proportion of harvested trees are not utilized for timber but as a source of fiber as wood pulp for paper or the production of boards, or to make rayon (viscose). In this case it is the cellulose that is wanted, the lignin that is present can cause discoloration and make a fiber more brittle though it is resistant to decay. However it is the lignin in other fibers that provides the strength and resistance to rot that has been utilized for millennia, twisted into twine or thread, and then woven into fabrics.

Lianas and climbers have long fibers and their strength is legendary, not just enabling Tarzan swung through the trees, but they are used to make everything up to rope bridges. Perhaps one of their most remarkable use is by the sky divers of the Solomon Islanders. Stems can be used directly as twine, plaited into rope or woven into basket weave.

Perhaps the earliest use of fibers from plants to make textiles was the production of tapa cloth beaten from plant tissue directly. These generally used the fibers present just below the skin or bark of the plant. The most famous is the tapa cloth of Polynesia and New Guinea obtained by soaking and pounding with mallets the bark of the paper mulberry (*Broussonetia papyrifera*). The technique was widespread: Otzi, the Neolithic man found preserved in a glacier on the border of Italy and Austria, was wearing a cape woven from the bast fibers of *Tilia* (linden or lime tree) woven with grass leaves.

Bark fibers were also used to make paper. From about 0-100 C.E. in China mulberry and hemp stems were pounded together to make paper. The Aztecs made paper from the inner bark of various species of fig (*Ficus*). The inner bark fibers were soaked in alkali and then beaten with a corrugated stone and glued together with vegetable gum. Finally the paper surface was sized with a chalky varnish.

A technology developed to separate out the fibers by retting, leaving the stems to rot, so that the lignified fibers which are resistant to rot remain. Retting is a fairly disgusting and smelly process. The simplest way to do this was dew retting. After harvesting the stems and drying them so that they died, the stems are laid out and allowed to soak up the dew. In flax this can take up to 3-6 weeks, during which time the stems are turned over several times. Alternatively after drying the stems are kept immersed in water. They have to be weighed down. The best contrivance is to use slowly flowing water to carry away the gels and mucilage released by the action of the rot causing micro-organisms. The smelly and hard part of the job is stripping the fibers from the remains of the plant. In jute this is sometimes done an individual stem at a time to produce high quality fibers, or by dashing bundles of stems backwards and forwards in the water. Flax fibers are obtained by a process called breaking, scutching and hackling: the stems are beaten with heavy wooden blades across a ridged break so that the core or boon falls away; the fibers are scraped to clean them; and then the long fibers are separated from the shorter ones. A similar breaking process was used to process hemp but lifting the heavy wooden boards to break the stem was unpopular work.

Hemp Cannabis sativa is one of the oldest crops cultivated for its fiber . Hemp is three times stronger than cotton and has a high durability but is not as soft as cotton. Its earliest use seems to have been in Central Asia and China. Prehistoric sites in China show the use of hempen cord, netting and cloth date from at least 5000 BC. Hempen textiles was used for clothing by all ranks until the development of silk, and even after that remained the garments of lower ranks. The fiber was probably taken west and south by Aryans south to India and by the Scythians west as far as Asia Minor. It was cultivated in Egypt by 1000 BC. It became a significant crop in Europe after the Roman period. It had a huge significance in the days of sailing vessels. It is reckoned that 80 tons of hemp were required for the rigging of a Tudor warship for example. It was also used to strengthen the canvas of sails. The supply of hemp was crucial for maritime powers. The rope produced in the Arsenale was famed for its quality. England got most of its supply from the Baltic states and Russia but tried to force American colonists to produce it.

Jute (*Corchorus*) provides a coarser fiber used in sacking (burlap) and carpeting. It has been grown in the Bengal area of India and Bangladesh from ancient times. Manufacture in the west began in the 1790s centered on Dundee in Scotland. The two species (*C. capsularis*, or white jute, and *C. olitorius*) grown for jute fiber are similar and differ only in the shape of their seed pods, growth habit, and fiber characteristics. The fibers are held together by gummy materials that are softened by retting for 10-30 days during which bacteria break them down. The fibers run the length of the stem, are loosened and jerked out of the stem. Kenaf (*Hibiscus cannabinus*), is used as an alternative to jute. The plant grows to a height of 5 m and provides fibers up to about 1 m long.

Flax (*Linum usitatissimum*) is the source of linen yarn. Linen is one of the oldest of all the plant based fabrics. Prehistoric Swiss Lake dwellers were using it and it was worn in Ancient Egypt. A linen industry was established in Britain by the Romans. The fiber is obtained by subjecting the stalks to soaking in water (retting), drying, crushing, and beating. The fiber strands measure about 30 to 75 cm. The yarn has strength, durability, and is resistant to attack by microorganisms and has a smooth surface that repels dirt. It also has a beautiful lustre. It absorbs and releases moisture quickly and so is comfortable to wear.

The Ancient Greeks used fibers from Spanish broom *Spartium junceum* for the sails of ships because they were resistant to salt-water . *Thymelea tartonraira* also provided strong fibers for sails and ropes and is now being used to make paper. Ramie (*Boehmeria nivea*) provides the longest, and most absorbent and silkiest of bast fibers but its fibers are difficult to free from the stem. The fibers have great strength if not bent and twisted. Pineapple provides another fiber.

Fibers from leaves are associated with the numerous vascular bundles in the veins. Monocot leaves are the most useful because of the long unbranched veins. The leaves of banana can be huge and hence the fibers can be up to 5m long and make ropes and a very strong paper, it is used for example to make tea bags, dollar bills and manila envelopes. *Musa textilis*, abaca or Manilla hemp, is the main species used in this way: In the Philippines the fibers are used to make clothing and in recent years they have been used to strengthen plastic components.

Papyrus was an early form of paper from *Cyperus* papyrus. Strips of the pith were laid at right-angles and pressed so that the sap glued them together. The plant also provided fiber for ropes and weaving. Boats were constructed from it. It also provided food from the lower parts and rhizomes boiled or roasted. Rice paper of the Orient was made by pounding sheets spirally cut from the pith of the rice paper plant (*Tetrapanax papyrifera*, a member of the Aralia Family (Araliaceae). The Aztecs and Mayans used the bark of a *Ficus* species to make a kind of paper not unlike papyrus.

Paper making was first perfected by the Chinese about 2000 years ago. Using the stem fibers of paper mulberry and hemp separated in water and floated they were allowed to settle on a mesh, before drying. Arab traders discovered paper in indo-China and it was through them that paper-making was introduced to Europe (Spain). Most paper today is made from the fibers of woody plants. The first step is to turn the wood into a pulpy mass. Several methods are used. In one process wood chips are cooked with bisulfites and then digested with strong acids. The softened fibers are then blown to separate them. Paper produced by this process has a tendency to discolor and become brittle with age. The alternative process is alkaline sulfate production using sodium sulfate, sodium sulfide and sodium hydroxide. The process also removes resins from conifer wood. Acid-free paper is normally used for books.

Sisal fiber is made from the leaves of *Agave sisalana*, a species of the agave family (Agavaceae) is the most important source. It is native to Central America, and has been used since pre-Columbian times. The fleshy and spiny lance-shaped leaves grow out in a dense rosette. Each is up to 2 m long and 20 cm across. About 300 leaves can be harvested at intervals from the plant. throughout the productive period. Outer leaves are cut off close to the stalk as they reach their full length. The fiber is usually obtained by crushing the leaf between rollers and scraping the resulting pulp from the fibers. Sisal fiber is coarse and inflexible but valued for ropes and twine because of its strength, durability, ability to stretch and resistance to deterioration in saltwater. Huge fortunes were made in Yucatan in exploiting it. Mereda was said to have more millionaires *Abaca* (Manila Hemp) comes from the leaves of a species of banana.

A range of fibers come from the fruits and seeds of plants. Coir is another seed-hair fiber but of a completely different character to cotton. It comes from the outer husk of the coconut, the (*Cocos nucifera*). The coarse, stiff, reddish brown fiber is made up of smaller lignified threads, each only up to 1 mm long. The processed fibers are up to 30 cm long and are light, elastic and highly resistant to abrasion. They are used to make brushes and matting as well as cordage. Kapok hairs are produced on the inner surface of the seed capsule of the kapok tree (*Ceiba pentandra*) a tall buttressed tropical tree species. Kapok hairs have a large empty lumen and have a waxy cuticle but are difficult to spin. Hence it is largely used as packing for mattresses and pillows as well as life-savers.

Cotton is the king of plant fibers. Seed is embedded in a mass of white unicellular hairs (trichomes) forming the boll inside the fruit capsule. The hairs (lint) are spun, twisted into usable thread which is tough and strong. Each hair may be up to 50 mm long. There are four different species of *Gossypium* cultivated. *G. hirsutum* (upland cotton) which originated in the Americas is the most important. It appears to be of a hybrid origin between *G. herbaceum* which is cultivated in Syria and Arabia and American species like *G. raimondii*. This happened naturally by the spread of a seed of *G. hirsutum* from Africa at least 5,500 years ago. Sea Island cotton has a higher quality lint with longer finer fibers appears to have originated as a hybrid between *G. hirsutum* and *G. barbadense*. It shares the quality with the high-quality Egyptian and Sudanese cotton grown today.

Cotton spinning and weaving technology first developed in the Sudanic Civilization in the region south of the Sahara as early as the sixth millennium B.C.E. in the region that *G. herbaceum* grows in the wild as a perennial shrub. Spindle whorls dating from this period have been discovered in northeastern Sudan. Interestingly this technology did not pass directly to the developing civilization in Egypt, they at first relied on linen textiles, but instead bypassed it and spread to the Middle East. A different species of cotton *G. arboreum* was being grown in the Indus valley between 2300 and 1760 B.C.E. Independently *G. barbadense* was being domesticated in South America; textiles dating from 3,600 B.C.E. have been discovered in archaeological digs from northern Chile. By 2,500 B.C.E. domestication was well underway and complex textiles were being woven. Meanwhile in Mexico by 3,500 B.C.E. *G. hirsutum* was being cultivated.

The cultivation of cotton is intimately associated with the continuation of slavery in the United States that was eventually to lead to the first modern war, the American Civil War between the industrialized Union and the slave-owning Confederacy. In a way cotton had a part to play in the huge disparity of culture and expectations of the combatants because it was the preparation of cotton, hugely labor-intensive that provided the motivation for the mechanization and industrialization of society. There had been an eightfold increase of slaves in the US between 1784 and 1861 to support in large part the cotton plantations.

Later the cultivation of cotton in the colonial lands was one motivation of western imperialist exploitation of places like the Sudan. The Gezira scheme established in 1926 brought irrigated cotton plantations to 800,000 ha of the land between the White and Blue Niles south of Khartoum, but turned subsistence farmers into transient wage laborers. Incidentally it also exposed them to much greater risk from malaria because of the greater areas of water in irrigation channels in which mosquitoes could breed. The first labor force was of black Americans imported like slaves from the southern states of the USA but they did not cope well with the conditions so that migrant workers, many from West Africa, introducing a further level of social inequality. Meanwhile local tenant farmers lost their ability to grow the crops of their own choice, had to buy their cotton seed and sell their cotton lint to the government run collective, and lost control over their own livings.

Ironically while the plantations relied on human labor and maintained the ancient social system of slavery into the modern age, it was the mechanized preparation of cotton textiles that drove the industrialization of the world. Richard Arkwright patented in 1769 a spinning frame that could spin weft and warp. The first mill as built in Nottingham and was powered by horses. It had 2,500 spindles controlled by only 50 operators and doing the work of 2,500 home spinsters. By 1790 the cotton mills were water powered and soon they were steam powered. In addition Eli Whitney's gin, patented in 1794, enabled a man to separate more than 20 kg of lint from the seed each day and not the 500 g possible by hand. After 1820 cotton overtook tobacco and sugar in value to world trade. The volume of production increased 10-fold between 1820 and 1860.

Assorted materials

Plants also a source of assorted materials from cork to thatch. In previous ages leaves of all sorts were used for wrapping and gourds as vessels. *Lagenaria siceraria*, the calabash or bottle gourd has a worldwide tropical distribution and was one of the earliest plant species to be domesticated separately. This appears to have happened separately in at least two places because it has different subspecies in Asia-Polynesia and Africa-America. Other gourds are produced from different genera in the same family, the Cucurbitaceae. Mosses that are scarcely used today except as a component of garden composts were used to wrap and keep food fresh, and in the absence of toilet paper for wiping bottoms. Many of these uses gave been superseded but the cork oak *Quercus suber* is still widely cultivated in the Mediterranean region for corks for wine bottles; though even this use is being subverted by plastic "corks" and screw-top bottles.

However many plants are still commercially important as the source of chemicals used in various kinds of industries. Oaks have been a prime source of tannins, though a wide variety of plants have been used. Beech and willow bark has also been used. The tannins are especially concentrated in the bark and insect galls. As their name indicates they were used for tanning leather. The polyphenolic tannins precipitate the proteins present in the animal hides. Tannins fall into two broad chemical classes, the condensed tannins, phlobotannins or catechols, and the hydrolysable tannins or pyrogallols. The condensed tannins They are polymers of flavan-3,4-diols (catechins) and flavan-3,4-diols (lueucoanthocyanins). *Acacia mearnsii* and *Schinopsis quebracho-colorado* have a high tannin content and bind very rapidly to the skin that is to be tanned but as a consequence they do not penetrate very deeply. *A. mearnsii* black wattle is from SE Australia but widely cultivated in the tropics. It has 60-65% tannin in the bark extract. They give the leather a reddish tan that deepens with age, but is not very light fast, and ultimately oxidation products weaken the leather.

Hydrolysable tannins are based on the molecule gallic acid or ellagic acid. They are obtained for tanning from several sumac (*Rhus*) species. In southern Europe, North Africa and the Middle East *R. coriaria* is used to tan sheepskins and to produce Cordoba and Moroccan leather; it has about 26% tannin content in its dried leaves. *R. hirta* from eastern North America is another useful source of tannin. Hydrolysable tannins give the leather a mellow greenish-brown color and other compounds present protect leather from the atmosphere. The dried fruits (32% tannin content) of the tropical *Terminalia chebula* (kashi or myrobalans) is another important source of a kind of hydrolysable tannin. It is also used in Ayurvedic medicine in India.

Turpentine is a mixture of volatile oils and non-volatile resins (rosin) extracted from different species of trees. Pines (*Pinus sylvestris*, *P. pinaster*, *P. palustris*, *P. caribea*) are the main source. The volatile part (turpentine) is used as a thinner for paints and varnishes and as a synthetic substrate for the production of a wide range of materials. Turpentine is only one of the range of products of pine used in wooden shipbuilding manufacture ("naval stores"). Canada balsam is the turpentine extracted from *Abies balsamea*. Rosin is used is the distilled residue of pine resin and is used to create friction between the bow strings and the instrument strings of stringed instruments.

Balsams are aromatic resins extracted from a range of plants, some of which have a medicinal value. *Myroxylon pereirae* provides balsam of Peru. Balm of Gilead and myrrh mentioned in the bible are sweet smelling resins used as incense and in cosmetics comes from small trees in the genus *Commiphora*: *C. gileadensis* native to the areas around the Red Sea. Camphor comes from *Ocotea*.

By heat and pressure resins polymerize and lose their volatile terpenes are transformed first to copal and at last become amber. The resin/copal is soluble in non-polar solvents and when these evaporate a shiny translucent layer is left. Copal is obtained from a number of other species. For example from *Copaifera* and *Hymenaea*, two related genera of African and American trees. *Trachylobium verrucosum* provides Zanzibar copal. *Guibortia demeusi* provides Congo copal. Kauri *Agathis australis* produces a resin called either kauri gum or as a semi-fossilized copal. Kauri gum was widely exploited in the 19th an early 20th century and hundreds of thousands of tonne of kauri copal were dug up in northern New Zealand for export and use in varnishes and in making linoleum. Gum diggers probed the soil with a spear to locate deposits. Linoleum was invented in England by Frederick Walton when he combined linseed oil, cork particles and kauri resin rolled to impregnate a fabric base. The Maoris also turned it into a kind of chewing gum by boiling it. *Agathis dammara* from the East Indies is another source. It gets its name from the dammars that are resins obtained mainly from species of Dipterocarpaceae. Lacquer is a similar product that is obtained by tapping the resin Chinese and Japanese plant *Rhus verniciflua*. The brilliant glass-like surface of lacquer work is produced by applying multiple coats of lacquer , allowing them to dry and polishing them on the addition of each layer. Mastic come from *Pistacia lentiscus* that grows in the Mediterranean region. It is used to make varnishes and glues but was also chewed as a gum.

Waxes differ from seed oil by being esters of simple alcohols and fatty acids and lack glycerol. Jojoba oil from *Simmondsia chinensis* lacks glycerol too and is really a liquid wax. It has the advantage of being tolerant of high temperatures so it can be used as a lubricant. It also penetrates the skin easily and so is used in cosmetics. Solid waxes include candelilla wax obtained by boiling the stems of Euphorbia antisyphilitica harvested from the wild in northern Mexico. It is a component of many furniture and leather polishes and in waterproofing materials. Carnauba wax is a hard wax obtained from the young leaves of *Copernica cerifera*, a type of palm from northeastern Brazil. These days carnauba is mainly used in shoe polish and car wax but previously it was used to make candles. Another source of plant wax is *Myrica cerifera* (way myrtle, candleberry or bayberry) from eastern USA; it makes pleasantly fragrant candles.

Water soluble gums include Gum Arabic is produced by wounding the branches of *Senegalia senegal* an species native to northern Sudan. It has many different uses. It provides glues and is added to lotions paints and inks. Perhaps most surprisingly it is added to many high sugar foods and drinks to prevent crystallization and fatty foods to emulsify them . It stabilizes the foam on beer. Another emulsifying gum is gum tragacanth from *Astragulus* species (mainly *A. gummifer*). It is used in pastes and spreads like mayonnaise and to thicken milkshakes. Gum karaya from *Sterculia urens* is relatively weakly soluble and a very sticky gum obtained from an Indian tree. It is also used in foods to thicken them as well as a dental glue and in hair-setting gels. Intermediate between gum

Arabic and gum karaya, but a better emulsifier than both is gum ghatti from another Indian tree called *Anogeissus latifolia*.

The gum-trees are the eucalypts. Oil of eucalyptus (eucalyptol) is a volatile terpene compound distilled from the leaves. It is used for flavorings, cough drops, and for the synthesis of menthol. Citronella another volatile terpene with a lemony fragrance that is supposed to prevent insects biting comes from the leaves of *E. citriodora*. Eucalypts are rich in kinotannic acid and used as tannins to convert animal hide into leather. One of the main Australian sources of kino is the common red gum (*Eucalyptus camaldulensis*).

True polysaccharide gums, come from sources such as the carob tree (*Ceratonia siliqua*) and chicle, a non-elastic rubber, a terpene gum from the latex sap of the sapodilla tree (*Manilkara zapota*). The latter is used in the soles of shoes and to make machine belts. It also provided the chewing gum of the Aztecs and modern Mexicans. Its modern use as chewing gum came from the experiments of Thomas Adams who, in 1867, received a supply of chicle arranged by Santa Anna. The victor of the Battle of the Alamo had fallen on hard times and was keen to raise money to finance his political resurgence by finding an alternative to rubber. Although Adams failed to vulcanize chicle he created a saleable product by adding sugar to chicle to make what we would recognize as chewing gum. Southeast Asian jelutong from the *Dyera costulata* is another source of chicle.

As well as oils for food some plants provide oils for industrial processes. Flax also provides linseed, an unsaturated oil used in the paint industry. Frederick Walton first produced linoleum in 1863 from oxidized linseed oil mixed with cork and pigments and pressed onto a jute or hemp backing. Castor oil is used in the production of nylon and other synthetic resins and fibers. It is converted to sebacic acid first. Purified wood cellulose is used in the manufacture of cellophane or rayon depending on the way in which it is extruded in manufacture. In the near future genetically manipulated crops will become important as sources of the raw materials of the plastics industry.

Rubber

Rubber can be made from the latex of many different species of plants. There are 15-18,000 species, in 20 different plant families that produce latex. Latex production is one of the most effective ways of discouraging herbivores. The latex is produced in special cells called laticifers. In some, called non-articulated laticifers the cell walls between the laticifer cells breakdown to produce multicellular, latex canals so that when the plant is damaged large quantities of latex bleed out. The laticifers are normally in fairly superficial and protective positions like the inner bark. Latex is a suspension of rubber particles along with, in different species, different toxins. Rubber is a complex hydrocarbon polymer with the ratio of carbon to hydrogen atoms of 5:8, called poly-isopyrene , though it was not until the 20th century that Hermann Staudinger proposed that it was a macromolecule and Wallace Carothers, a chemist at DuPont and the inventor of nylon, demonstrated it. The important latex producing families are Apocynaceae, Asclepiadaceae, Asteraceae, Euphorbiaceae, Moraceae and Sapotaceae.

Latex was tapped and utilized in Central America at least from the time of the Olmecs. Footwear and vessels were made and also the ball was used in the ritual ball game, called tlachtli in Aztec that was played throughout Central America and on the main Caribbean Islands; Columbus observed the balls on his return to Hispaniola. Opposing team attempted to hit the ball to their opponents' end of a large "I –shaped" courtyard with sloping sides, and through a stone hoop, but without using their feet or hands, no mean feat because the ball was bowling ball-sized and weighed about 2 kg. The game was the occasion of heavy gambling, but also on occasion at least represented a battle between gods, with the losing team captain being beheaded in sacrifice.

The Central Americans had a way of producing utilizable objects by using the addition of the *Ipomaea* juice to latex from *Castilla elastica*, a species from the fig family Moraceae. After addition and stirring for fifteen minutes the latex turns into a solid mass than can be shaped into a ball by hand . For Europeans the problem with latex is that it rapidly congeals and so has to be worked immediately. In addition the rubber produced is sticky when hot and brittle when cold. These qualities restricted the use of rubber in Europe; it was used " in the 18th century at first to make erasers to rub out pencil marks, hence it was called "rubber. It was not until techniques were discovered to re-dissolve the rubber and to strengthen it through vulcanization that its use became much more practicable. The first major step was when Charles Macintosh found he could dissolve India rubber from *Ficus elastica* by soaking it in a bath of naphtha. In 1825 he opened a factory in Glasgow to use the rubber solution to coat textiles, and steam to evaporate the solvent, to make waterproof cloth, hence we call a particular kind of raincoat a Mackintosh. Unfortunately the rubberized textile was still likely to either melt or go hard and crack.

After Mackintosh died this first rubber manufactory was taken over by a man called Thomas Hancock who had the business acumen to patent in 1844 the discovery of Charles Goodyear in America of vulcanization. The application of sulfur and steam produced a rubber that was resistant to changes in temperature and was elastic. Vulcanization is now carried out with the addition of an accelerator like zinc oxide. It works by cross-linking the poly-isoprene chains. The rubber ball had been reinvented! But now a wide range of additional products were produced. The market for rubber was given a boost in 1888 when John Dunlop invented pneumatic bicycle tires. In 1895, André Michelin made the first pneumatic car tires, and in 1904 P.W. Litchfield of the Goodyear tire company patented a tubeless tire. However that it was not until 1911 that car tires, ones with a rubber inner tube, went into production successfully.

A number of different species were tried as a source of latex. At the end of the 19th century and the first decade of the 20th century the rubber trade was dominated by latex collected from wild *Hevea* trees and also from other species. The most important were in the Americas Panama rubber *Castilla elastica* from Central America and ulé or uli *C. ulei,* from the Amazon. It was balls of *C. elastica* rubber that Columbus observed. In Central Africa the latex of the vine *Landolphia kirkii* (Apocyanaceae) was exploited, especially in the fiefdom of Leopold II in the "Free" Congo. Several other species have been used to provide latex for rubber but none has really become commercially viable except when rubber has been in short supply. *Gutta percha, Palaquium gutta,* from the family Sapotaceae, provides a rubber that is more brittle. Another species from this family, *Manilkara bidentate,* provides balata rubber which is also non-elastic and used to make machine belting.

The most important source of latex is from pararubber, *Hevea brasiliensis.* It was exploited exclusively from the wild forests of

South America until it was introduced to South East Asia. It was used to provide rubber but also the seeds were eaten. H. A. Wickham obtained 70,000 seeds and 2,800 seedlings were raised at Kew. The first plantation was established in Sri Lanka in 1872. By chance it was introduced without South American Leaf Blight *(Microcyclus ulei)*. A few plants sent to Singapore, Malaysia and Indonesia were used to establish commercial plantation there, first begun in the 1890's by Ridley. However it was not until methods were introduced to repeatedly scrape the bark to maintain the flow of latex that sian rubber out-competed that available from America and Africa. Today it is estimated that there are 7,322,792 Mha of rubber plantations.

The pararubber family, the Euphorbiaceae, is a large and diverse family that includes trees and in Africa succulents that look strikingly like cacti, but are readily separable from them because they produce latex and cacti do not; cacti are also only found naturally in the Americas. The largest genus in the family, and the second largest genus of all plants is Euphorbia; it includes a range of species of which have also been exploited for their latex. The latex of *E. balsaminifera* was used as a bird-lime to trap birds in its stickiness. *E. intisy* from Madagascar was a source of latex for intisy rubber. Even the latex of poinsettia, *E. pulcherrima* has been used as a depilatory in Mexico. The *Euphorbia* chemical factory also provides a wax, from *E. antisyphilitica* (candelilla) from south-western North America, that is added to polishes, added to electrical insulation, and used for waterproofing. Another species E. tirucalli is a useful fuel, because of its high hydrocarbon content , and in India as charcoal is added to fireworks.

Dyes

A variety of plants, even commonplace ones provide dyes . Some have been given a common name recording this use: Dyer's Greenweed, *Genista tinctoria*, provides a yellow dye or Kendall Green when mixed with woad. Dyer's rocket or Weld *Reseda luteola* provides another yellow. There are hundreds of dye providing plants yellow from sage, tansy, yarrow, coreopsis and goldenrod flowers or onion skins, green from foxglove, rosemary, lily of the valley or rhododendron leaves or Queen Anne's Lace, purple from blackberries, geranium or lady's bedstraw, red from dandelion, Potentilla or St John's wort, brown from burdock, fennel, poplar, acorns or marigold flowers, blue from elder.

What are really valued are those dyes which are fast and long-lasting . Dyes are of three sorts; "substantive" or direct dyes, and vat dyes, that both require no mordanting to adhere to the fiber, and "adjective", "natural" or mordant dyes that require mordanting. Mordants like alum (aluminium sulfate) allow the dye to penetrate the fiber and holdfast. The name comes from the Latin *mordere* for "to bite". Plant fibers like cotton and linen are harder to dye than wool or silk and mordants help. Mordants are normally metal ions. Alum is commonly used because it has relatively little effect on the color but mordants like ferrous sulfate or tin chloride darken or brighten the color respectively. Tannic acid, from example Brazilwood sawdust, is a good mordant for plant fibers but it tends to tan or brown the color.

Vat dyes are applied in a chemically reduced and soluble "leuco" form (leyko in Greek means colorless or white), made by heating them with alkali, and then oxidized chemically, or just by being exposed to the air, to reform the insoluble pigment thereby fixing it to the textile (or skin!). The leuco form of vat dyes is often a complimentary color to the oxidized dye. Indigo, indigotine, the premier vat dye, is a deep-blue and fast dye. The importance of indigo is that it was the only blue dye until the introduction of Prussian blue in 1810. Tyrian or Imperial purple came from a mollusk but could be obtained by adding a red dye like madder to indigo. Similarly Lincoln green was produced by overdyeing blue woad with yellow from Weld or Dyer's Greenweed.

Indigo has an ancient use. Indigo's name comes from the Greek Indikon a reference to its source in Ancient times, India from the plant *Indigofera tinctoria*. By the second millennium B.C.E. Indians had developed a sophisticated dyeing technology to go with their domesticated cotton . Indian textiles were traded to Mesopotamia and by the first century CE were being traded in Egypt. The dominance of Indian textiles remained unchallenged until Britain's industrial revolution.

In the Americas two other species of *Indigofera, I. suffruticosa, I. arrecta* provided indigo. An early use of indigo is described from a Babylonian tablet of the 7th century BC: uqnatu a lapis colored wool was produced by repeated dipping and air-drying of the cloth.

Indigo can be obtained from about 200 species of plants. In more northerly regions there were alternative sources of indigo from woad *Isatis tinctoria* and dyer's knotweed *Polygonum tinctoria*. The fashion followed by Ancient Britons of painting the body with indigo was described by Caesar when he invaded Britain and it may even have given Britain its name from the Celtic language word "Brith" meaning paint or mottled. The Ancient Britons probably dissolved the indigo in stale urine which is slightly alkaline the after they had painted their bodies with the dissolved leucoindigo which is yellow; like magic their bodies turned blue when exposed to the air.

In Japan there was a different technology involving the boiling of indigo with thermophilic anaerobic bacteria that produced hydrogen converting it to soluble indigo white . Indigo was widely used, even to dye the first blue uniforms of British policemen. Today it is most widely used to dye denim. Several methods were devised to produce patterned cloth. Pencil blue was produced using dye with added arsenic trisulfide and thickener. Paler china blue patterns were produced by printing with insoluble indigo and then immersing the cloth in a sequence of baths of iron(II) sulfate to oxidize it.

Like many vat dyes indigo is based on an organic ring structure, in the case of indigo it is a rather beautiful four ring structure with rotational symmetry. Research towards the development of an artificial indigo started in 1870 by BASF in Germany, originally a supplier of natural indigo, was momentous for the development of chemical industry; Adolph von Baeyer synthesized it in 1880, described its structure in 1883 and by 1894 the technique for making artificial indigo on an industrial scale had been perfected and the organic chemistry industry had been established. Baeyer was awarded the Nobel prize for chemistry in 1905.

Over 95% of natural red dyes are anthroquinones which have a three ring structure. Two of the most important cochineal and lac come from animals but alizarin is a plant dye. *Rubia tinctoria* madder and two other *Rubia* species, *R. peregrina* wild madder or majithro and *R. cordifolia* Indian madder are also sources. *Morinda citrifolia*, the Indian mulberry also yields alizarin. *Rubia tinctoria* is said by some to include its own mordant but faster stronger dyes are made by adding tin, alum or copper. Anthroquinones are also present in buckthorn species and anthraquinones are the source of several dyes such as Chinese green indigo used to dye silk from *Rhamnus utilis*.

The napthoquinones have a two ring structure. They are direct or substantive dyes and mostly brown or purple-grey. Henna (*Lawsonia inermis*) is used to dye the body and hair all the way from North Africa and to South Asia. It has a very ancient use and was used, for example, by Roman matrons to hide their grey hair. Its main dye component is lawsone. It is a very fast dye on animal based materials such as hair and skin and can also be utilized to dye wool and silk and leather but not cotton. Like other naphthoquinone direct dyes, such as those from walnut hulls and safflower it does not react with the cellulose of the plant fiber and therefore lacks fastness. Juglone is obtained from a variety of walnuts *Juglans*. Confederate gray is obtained from butternuts (*Juglans cinerea*). They may continue to bleed every time the textile is washed. However they are generally light-fast.

Red dyes come from the roots of several species in the borage family Boraginaceae; alkanet *Alkanna lehmannii* is a red dye used to tint port, thermometer fluids and stains fats. Others are *Anchusa officinalis* and *Onosma echiodes*. Pitti a red purple dye comes from the roots of *Ventilago madras*.

Flavones are three ring chemicals but unlike anthroquinones one of the rings is linked separately from the other two. Flavone dyes can be used as direct or substantive dyes but are much improved by mordanting. They provide mainly yellows and browns. Weld *Reseda luteola* produces a yellow dye called luteolin from the flowers and stalks. Quercitron a yellow powder produced by grinding the inner bark of the oak *Quercus velutina*, dissolved and treated with acids produces a precipitate quercetin dye. Quercetin is also found in Persian berries (*Rhamnus infectoria*) and buckwheat leaves (*Fagopyrum esculentum*). Quercetin is fast in cold water but a little un-fast in hot water and dissolves in alcohol. A more fast dye is "Old fustic" comes from *Chlorophora tinctoria*, (= *Maclura tinctoria*) provides old fustic. A related species osage orange (*Maclura pomifera*) produces a bright yellow dye from the heart of the wood. Other flavone derived dyes are fisetin from young fustic or Zante fustic (*Rhus cotinus* or Venetian sumach), chrysin from poplar buds (*Populus*), apigenin from parsley and galangin from galangal root (*Alpinea galangal*) and myricetin from the podocarp conifer *Nageia nagi*. Other flavone dyes are from the yellow flowers of chamomile (*Anthemis nobilis*) and marigold *Tagetes erecta, T. patula* gold and orange for dying wool and silk, tesu, tisso or keshu, the golden yellow flowers of the Indian Flame of the Forest, *Butea frondrosa, B. monosperma*, said to be one of the most beautiful of all flowering trees, and dolu, the brown yellow roots of *Rheum emodi*, a kind of rhubarb.

The brown dye cutch comes the condensed tannins, based on flavones including quercetin, found in the heart of the wood of *Senegalia catechu* (betel nut tree) and *S. chundra*. Betel nut wood was used to produce true khaki (from the Hindustani khak for dust) obtained by boiling up cotton with sawdust from the heartwood. Khaki was first adopted by Indian troops from about 1846; the first were Harry Lumsden's Corps of Guides. A whole range of other substances were also used tea, coffee, mulberry, curry powder, tobacco and even mud. The camouflage value of khaki took a while to catch on; the last battle fought by British redcoats was on the Egyptian-Sudanese border in 1885. Khaki was adopted for all overseas troops in 1897 and home troops in 1902.

Chemically different yellow and orange dyes are an iso-quinoline dye that comes from barberry *Berberis vilagaris, B. aristata* and a chromene dye from kamala, the monkey-face tree *Mallotus phillipensis*.

A more significant yellow-orange color comes from saffron, a carotenoid pigment called crocetin. To prepare 1kg of dried saffron the stigmas from 100,000 flowers must be collected. Saffron, called Kesar in the Kashmir, has always been a rare and valuable dye. It comes from *Crocus sativus* and is obtained from the stigmas. The plant grows from Spain to the Himalayas but seems to be often sterile and hence must be considered a domesticated species. Saffron robes represent the adoption of an ascetic life for Jains, Sikhs and Buddhists. In Thailand 300,000 men wear the robes as a sign they have adopted full Buddhist monastic ordination. This religious identification, which may arise from its golden color, was first made by the Hindus who still used it to dye robes and to make the tilaka mark on the forehead. This may represent the mystic third eye. The sacred cow Khamadenu has horns stained with saffron. For Sikhs and Hindus saffron is also signifies the fight against injustice and is commonly used on flags. Saffron was also identified with the Cretan goddess Eos or Aurora, the goddess of dawn who had saffron dyed fingers. Homer in the Iliad describes her,

"Dawn in her saffron robe rose from the River of Ocean to bring daylight to the immortals and to men".

Safflower *Carthamnus tinctorius* provides an orange dye. Safflower may have got its name from saffron but it was much more available. *Carthamnus* comes from the Arabic *quartum* for the color. It was used to dye carpets in Iran and Afghanistan. It is a particularly effective direct "substantive" dye made soluble by dissolving it in acidic solution but in Japan dyed silk fabric was treated with alkali (wood ash water) and then acid (sushi su - sake that had turned to vinegar) to produce a pink color called kurenai, that was reserved for the Empress and women of high rank and forbidden to the masses.

The inky purple/black hematein dye that comes from logwood or campeche (*Haematoxylum campechianum*) a tree that grows in Central and South America is a benzophyrone. It has an affinity for acidic material and is used in microscope preparations to stain for example, nuclei.

When westerners traveled to the Pacific Islands they were impressed by the tattooing they observed. A blue pigment prepared from the baked nuts of the candlenut tree *Aleurites moluccana*. The fashion was adopted by sailors and exported to the west. However, many dyes used by tattooists today are industrially produced dyes used in paints and inks.

Bodily pigmentation was also carried out using Bloodroot (*Sanguinaria canadensis*) that like many other members of the poppy family (Papaveraceae) has a colored latex, used as a red dye by native Americans to paint their skin and to dye fabrics. Its red color had many connotations with blood and so it was widely used medicinally, but it was also associated with fertility and used as an aphrodisiac. One story relates that men of a certain tribe would paint their hand with bloodroot and shake hands with a woman they wanted to have sex with. Its use for adornment, along with the application of red ochres, may have given rise to the term "redskin".

Many plant dyes, used to dye textiles, have been supplanted by chemically produced aniline dyes. Such is the case with alizarin. However concern for the use of natural products as food additives has maintained the use of foodstuff dyes that come from plants. These are mainly substantive or direct dyes like annatto. *Bixa orellana* is the source of annatto (mainly a carotenoid called bixin) the yellow dye utilized in many food products like margarine. Crocetin the saffron dye is a related carotenoid that is also used as a flavoring. The spice turmeric (*Curcuma longa*) has increasing use as a yellow food colorant from a napthoquinone called curcumin.

An interesting question is what these dyes do in nature? What are they for? Obviously when they come from flowers they are showy and attract pollinators but in the heartwood or bark it seem their color is accidental and they act mainly as chemical deterrents against wood-boring insects and as fungicides. In the roots they may act as allelopathic agents preventing the spread of the roots of other plants. For example juglone is a phytotoxic poison.

As well as being dyed with plant dyes textiles were also bleached with plant extracts. Linen was bleached with poppy juice.

Fuel

Throughout human history wood has provided the primary source of fuel. It is only in the last few hundred years that fossil fuels have provided an alternative. The industrial scale use of wood for fuel came with the great civilizations. Roman towns had an insatiable requirement for fuel. The woods of southern England were coppiced to provide a continuous supply of wood for Roman industry such as the iron workings of the Weald and the Forest of Dean. The extent of the Roman workings in the Weald suggest about 23,000 acres of coppice wood would have provided the need. The distinction between timber and wood is recognized in the Latin words *meremium* (= timber) and *boscus* (= wood) to describe it. *Carpinus betulus* (Hornbeam) was a favored fuel tree and became more widespread in the Roman period. Hornbeam was one of the best woods for this purpose. The iron working industries of the Forest of Dean and elsewhere were based on extensive coppice systems. The requirement for charcoal maintained many coppices. Even in 1905 a third of all woodland in England was coppice. A trade in fuel wood was established.

Charcoal was produced by "colyers" working in the forest. Wood was cut to the correct length and then stacked to dry for several months. Then it was arranged in a circle about 24ft in diameter around a central pole and piled to make a tall cone. The outside was covered with ash and bracken to make an airtight seal. Then the central pole was removed and glowing embers added to ignite the cone. It was left for two to four days. *Carpinus betulus* (Hornbeam) was the preferred wood for charcoal but *Quercus*, *Fagus* and *Betula* were also used. Coppices of hornbeam were still supplying London with much of its fuel in Tudor times. Coppicing for charcoal became less economic after the railway network made available cheap coal from the north. Coal was brought by ship to London from Newcastle in the Tudor era.

Wood is still a vital fuel in many parts of the third world. In the semi-arid areas of Africa species of acacia provide a major source. These trees and bushes have an important place in the delicate ecology of areas sensitive to desertification.

Scents and perfumes

The word perfume comes for the Latin for through smoke, *per fumen*, a reference to the use of incense in religious ceremonies. They were the most valuable of products. The wise men brought to Jesus presents of frankincense and myrrh, scented resins from the tree *Boswellia sacra* and thorn-bush *Commiphora abyssinica*. Frankincense was used as an incense in religious ceremonies and therefore represents Jesus' divinity. Myrrh was used in embalming, its scent last for many decades and it preserves the flesh; it represents the humanity of Jesus. Myrrh and frankincense is still used in incense, but more familiar in joss sticks or agarbatti (incense around a stick) and dhoop sticks and cones (pure incense). Popular scents are sandlewood, vetivert and patchouli. Sandlewood (*Santalum album*) is a tree with scented wood that is used in Hindu and Buddhist cremations. Vetiver (*Vetiveria zizanoides*) is a grass with scented roots. Both are from India as is patchouli, from the herb *Pogostemon cablin* and other *Pogostemon* species. Patchouli became the most redolent scent of the hippy era. Ubiquitous at festivals, concerts and in boutiques from the 1960s onwards. It was strong enough to mask to some extent the smell of cannabis smoke.

It is commonplace how smells can trigger a memory. They can alter our mood and we fall in love with the smell of a loved one. They are pheromones and aphrodisiacs. Humans have turned to plants to expand the range of available scents. Many scents come from flowers. The same scented flowers that were cultivated by the Ptolemaic Egyptians are favored in gardens today, jasmine, rose, lily-of-the-valley, stock. Flowers were grown in gardens in the Roman Empire for garlands and for perfume and there was an extensive luxury trade in flowers. Egypt exported fresh flowers to Rome! It was not until Avicenna in the 10th century distilled the scent of Damask rose that lasting perfumes were obtained. The scents of perfumes, and also many herbs and spices, comes from volatile oils, sometimes called the atar, attar or otto, a word which probably has an Arabic root. The atar of roses, for example, is mainly the result of two essential oils; geraniol and 2-phenylethanol. Today perfumes are regarded as scents with 22% or more essential oils and other scents have successively less; Eau de Parfum 15 -22%, Eau de Toilette 8- 15%, Eau de Cologne about 4%, and Eau Fraiche 1-3% essential oils.

Lavender *Lavandula* was widely used by the Romans, to provide scented water to wash, or placed with clothes and textiles to scent them Its name is said to come from the Latin *lavare* to wash. There are several highly scented species several with a Mediterranean and Near Eastern origin. *L. angustifolia*, English lavender, gets its vernacular name from an especially sweet-smelling variety of *L. angustifolia* that was introduced to England in the early 1600s, and became the mainstay of the English industry, though lavender was probably introduced much earlier by the Romans and was a mainstay of medieval gardens. The first commercially produced scent in the British Isles associated with particular places like Mitcham (Surrey), Hitchin (Hertfordshire), Market Deeping (Lincolnshire). Plants are half cut after 4 years and 3,500 plants provides about seven liters of otto. In lavender water it is often mixed with as diverse scents as those of jasmine, rose, orange blossom, canella bark and wallflowers. *L. stoechas* is French lavender but in the western Mediterranean region, in regions like Provence, commercial production is based on a hybrid between *L. angustifolia* and *L. latifolia* (*L. x intermedia*), called Lavandin.

An ancient technique for extracting scents was enfleurage, a kind of which was being practiced by the Ancient Egyptians. The scented petals are laid out on an unscented layer of plant or animal oil or fat, such as pig lard, and the oils allowed to diffuse out, and then replaced daily by fresh petals, until the oil or fat is heavily impregnated. This highly labor intensive method enabled the extraction of delicate fragrances that might be destroyed by heat. The scented pomade produced could be used directly or the essential oils in it extracted with alcohol.

The alternative technique of distillation involves the passing of steam through the flowers so that it becomes laden with volatile oils and then allowing it to condense. The insoluble oil is then drawn off. 2,000 roses yield just 1 gm of the attar of roses. The technique of distillation was invented in the East. By the second millennium distillation was being used in the Middle East to create perfumes by concentrating the essential oils of various plants . Knowledge of the technique was preserved here in the Middle East and Egypt became a center of the technology. There were important centers in Damascus, Sabūr and Jūr in Iran and Kūfa in Iraq and perfumes were exported widely, as far as China. Al-Kindi produced a "Book of Perfume Chemistry and Distillation) in the 9th century C.E. .

Various parts of the plant were used. The process today involves either cold pressed extraction, extraction with ether, or steam distillation and then washing and dilution with alcohol. Roses, violets, jasmine, tuberose, mimosa, jonquil, and orange-blossom and bergamot from the Mediterranean region and lavender and peppermint from more northern regions are the most important sources of scented essential oils. Alcoholic extracts are complex mixtures. Bergamot from *Citrus bergamia* contains α-pinene, β-pinene, myrcene, limonene, α-bergaptene, β-bisabolene, linalool, linalyl acetate, nerol, neryl acetate, geraniol, gerianiol acetate and α-terpineol.

In an age when bathing was not frequent huge amounts were spent on perfumes. Elizabeth I spent £40 for perfumes from one supplier. But perfumery was much more important than just a means of making life smell more pleasant. From the Late Middle Ages there were regular episodes of plague. For example major plague epidemics starting in Germany and the Netherlands came in the years 1498, 1535, 1543, 1563, 1589, 1603, 1625 and 1636. There was a widespread belief in the link between bad air and disease. Perfumes were seen as a prophylactic against disease. A posy of scented flowers was held to the nose to keep away the evil-scented are and in times of plagues some people took to wearing masks with long pointed snouts filled with scented herbs, without much success as the nursery rhyme points out

Ring a ring o' roses,
A pocket full of posies,
Hatishoo! Hatishoo!
We all fall down.

The ring of roses was the first marks of the infection on the skin and the sneezing a sign of the virulent pneumonic form of the disease.

The production of scents moved into Europe via the Moorish influence in places like Toledo in Spain and Sicily. Often it was Jewish scholars who translated Arabic works into Latin. Michael Scot (c. 1175-1232 C.E.), and Andrew the Jew, translated Avicenna into Latin. In Sicily the Jew Farraguth translated other Arabic works. Spanish perfumed leather was famous. It was produced by soaking in an otto composed of the oils of neroli rose, sandalwood, lavender and verbena with small quantities of oils of clove and cinnamon spirit, plus gums of benzoin, before the skins were rubbed with gum tragacanth and musk pressed together. By the time of the renaissance perfumery was well established in Italy and even in northern Europe.

With the opening up of the world the tropics provided a source of a range of new and heady scents to Europe. According to one story Frangipani was a scent first developed by the Frangipani family in Rome based on orris root, civet, musk and spices. In the West Indies a tree was found with a similar scent (*Plumeria alba*) and named Frangipani, after Mercutio Frangipani who had accompanied Columbus to the West Indies in 1492. Its Latin name was for the French monk botanist Father Charles Plumier who died in 1706.

Ylang-ylang (*Cananga odorata*) from the Philippines provides one of the top notes of Chanel No. 5., invented by Earnest Beaux for

Coco Chanel in 1921, but on its own is rather overpowering. The terminology used to describe perfumes shares the same obscurity as those used to describe the flavors of wines. In Chanel No. 5 ylang it is combined with neroli or Seville orange (*Citrus aurantium*), as the other top note with a heart of jasmine (*Jasminium grandiflorum* or *J. odoratissimum*), rose (*Rosa damascena*) and a woody base of sandalwood and vetiver.

However the modern perfume industry did not start with the launch of Chanel No. 5. but much earlier perhaps with Eau De Cologne. It was celebrity endorsement that launched it, in this case the endorsement of Madame du Barry, mistress of Louis XV, and later by Napoleon, who is said to have used bottles of it. An Italian barber invented it 1709 as Aqua Admiralis, a perfumed water, with of grape alcohol, neroli, bergamot, lavender and rosemary. He moved to Cologne where it was at first sold in apothecaries. It was popularized across Europe as a cure-all in the period of the Seven Years War. It was the French that first called it Eau De Cologne. Celebrity endorsement is now almost an absolute requirement for the launch of a new perfume. Perfumes are now part of the fashion industry packaged in extraordinary bottles and selling sex through gimmicky advertising. There have been fashions for different kinds of scent, almost all of plant origin whether they are musky or fresh, with lily or vanilla top notes.

5 Your Sweetness Is My Weakness

For millennia humanity has sought out the sweetness of sugar and sought to release its demon alcohol. Sugars are made up of a simple unit, normally formed in a ring called a monosaccharide with six carbon atoms, six oxygen atoms and twelve hydrogen atoms. Six of the hydrogen atoms are linked to the oxygen atoms as hydroxyl groups and it is the different orientation of these that determines what kind of monosaccharide it is, such as glucose (sometimes called grape sugar – its name comes from the Greek for sweet wine), fructose (sometimes called fruit sugar – its name comes from the Latin word for fruit) are just two. Fructose is the major component of corn syrup. Table or cane sugar is a disaccharide called sucrose and is composed of a unit of fructose and a unit of glucose linked together. Maltose is a disaccharide made up of two glucose units. It is found in germinating grains. Fructose is slightly sweeter than sucrose, which is sweeter than glucose. Maltose is the least sweet. Starch is made up of many glucose molecules linked together. It is glucose that is converted by the process of fermentation by yeasts into ethanol (ethyl alcohol). Luckily disaccharides and starch readily break down, hydrolyze, into their basic monosaccharides in acid conditions and especially when enzymes like invertase or amylase are present. Three different enzymes are used to produce the high fructose corn syrup from maize starch; alpha-amylase is used first to produce short chain polysaccharides, then glocosamylase to produce glucose and finally glucose-isomerase converts the glucose to fructose. All this is now done in vast with enzymes from bacteria, mainly Bacillus, and the fungus Aspergillus.

Honey

"Sugar, ah honey, honey, You are my candy girl" so sang the Archies and honey has always been one pure source of sweetness – plant derived but collected for us by bees. Nectar, with pollen, the

raw component of honey, is a mixture of sugars, mostly fructose and glucose. Typically a honey might contain 38% fructose, 31% glucose, 1% sucrose, 9% other sugars, with the rest mainly made up of water. Nectar sugar concentration varies between about a quarter and three-quarters by weight. Some large families of flowers like the cabbage, carrot and daisy families produce a nectar which has low sucrose but high fructose and glucose composition but some plants like *Berberis* and *Helleborus* produce a nectar which has mainly sucrose. Others produced a balanced menu with the different sugars in roughly equal proportion. *Abutilon* is like this. Some plants even produce a toxic nectar. This is especially common in the heather family. Metheglin made in Pennsylvania in the 1700s from the honey of *Phyllodoce breweri* was high in these toxins and in classical times the soldiers of Xenophon, two days march from Trebizond in the Caucasus, gorging themselves on honey made from native azaleas and rhododendrons (*R. ponticum* and R. luteum), vomited or had diarrhea, or behaved as if they were drunk and passed out for three days . (Xenophon's Anabasis 4.8.19). Aristotle and Dioscorides called this mad-making honey from Pontus *mainomenon*.

Bees tend to favor nectars relatively rich in sucrose but the enzyme invertase supplied by the bee breaks this down into its constituent fructose and glucose in the honey . The oldest record of keeping bees in hives is by the Egyptians more than 4,000 years ago but cave paintings depicting the collection of honey pre-date this by many millennia. There was a remarkable understanding of bee society as shown by the Greek myth that Zeus was fed on the honey of queen bees during his upbringing, although they also thought nectar was extracted from the air and wax from pollen. Ambrosia and nectar (its liquid form) were the food of the gods and probably made from nectar. Honey was offered as a tribute to the gods in both Egyptian and Greek custom. Although the honey bee is not native to the Americas there is evidence of Mayans providing hives for other species of social honey storing bees.

Melomels are fermented fruit juices with added honey; piment from grape juice, cyser from apple juice, perry from pear juice, morath from mulberries

Sugar

Barry White sang:

"You're a lovely sight, lovely sight to see, mmm, hmm
The way you give me your sweet love
Any place and time when you look at me I get weak in my knees
I'm so thankful that you're mine
You sweetness is my weakness, yeah, yeah"

Alternatives to honey were slow to be established. They relied on the development of a technology to extract the sugar from the cane or root. Sugarcane probably originated as a crop in New Guinea, one of those forest crops exploited from the wild. As sweeter, juicier and more easily chewed canes were discovered they were gradually taken into cultivation and domesticated as early as 6000 B.C.E. . In the process *Saccharum robustum* the wild species was domesticated as *S. officinarum*, the noble canes. Plants are propagated clonally, by pieces of stem and flowering is selected against because it checks growth, and yet major evolutionary changes arising from hybridization, and modern plant breeding, have required and require flowering plants. Cultivated clones tend to flower least near the equator.

The cultivation of *S. officinarum* spread both north and west so that even by 3000 years ago it was chewed for its sweetness in India and China. In these middle latitudes flowering occurred more regularly and hybridization occurred with a widespread and variable wild species called *S. spontaneum*. The thin canes, *S. sinense* that thrived in the monsoon climate, and are still cultivated in India and southeast Asia for sugar production but also for fodder were the result. Plant breeding in the 1920's backcrossed *S. sinense* to *S. spontaneum* to produce a genetic base from which the modern nobilized canes were selected. Thousands of cultivars have been named.

Sugar cane has had an influence on human history that rivals that of the grain crops. A variety called puri was first extracted to provide refined sugar in India. The canes are pressed to squeeze out the sweet sap. The word sugar is derived from the Prakit Sacchari. Prakit is a language with a Dravidian origin and it was the

Dravidians who were making a treacle called guda by 1000 B.C.E. utilizing a machine to extract the juice from the canes. Juggeri, jaggery, panela or gur is the unrefined whole sugar derived by allowing the treacle to evaporate, helped by boiling, a technology well established by the time of the Gupta kings in 350 B.C.E. Sugar with Ghee was carried by South Asian sailors. Sugar cane cultivation and technology spread westward, at first following the invasion of India by Darius in 510 B.C.E., then Alexander the Great in 325 B.C.E., and later more notably following the spread of Islamic culture. "The reed that gives honey without bees" as it was described by the Persians, or "the sacred reed" as it was described by Alexander came to be cultivated throughout the caliphate along the Nile, and after the development of better irrigation technologies allowed it to be cultivated Syria, Lebanon and Palestine and the Mahgreb reaching Spain by the 8th century C.E.

Refined cane-sugar, sucrose, was a valuable commodity; in the 14th century it was worth 10 times more than honey. In England it was the preserve of the royal kitchen but by the seventeenth century sugar had ceased to be a luxury and had become a staple. "For every ton consumed in England in 1600, 10 tons were consumed in 1700 and 150 tons in 1800" . By that time sugar consumption was about 5 kilos per head.

It started in the early 15th century with the cultivation of sugar cane in the colonies established by the Portuguese on Madeira and the Azores and Spanish on the Canaries. In 1452 the first sugar-mill and plantation was established on Madeira: in a hundred years there were 40 mills and 3,000 slaves. After the conquest of the indigenous Guanche people in the 1490s sugar plantations were established in the Canaries. From there it was a short step to its introduction into the West Indies and the Americas encouraged by the loss of other sources in N. Africa and the Near East because of the expansion of the Ottoman Empire. Columbus introduced it to San Domingo in 1493. It flourished in the deep soils and the warm wet climate of the West Indies. By 1530 there were already 28 sugar mills on the island. The Portuguese established production in Brazil but the sugar economy transformed the Caribbean and for the next 300 years Spain, France, Holland and England competed and fought for the "honey-pot" islands. The Dutch eventually failed to take over the Portuguese sugar plantations in Brazil but they financed England's involvement, which began when sugar was planted on Barbados in 1643. Soon four-fifths of the island was covered by sugar-cane plantations. Other islands were occupied. Jamaica was taken from Spain in 1655. By the end of the century it was reckoned that an investment of £5000 in a sugar plantation returned £1000 annually. The refining of the sugar was reserved to the homeland and the government took its cut. Import duties raised £280,000 between 1699 and 1701. The sugar economy underpinned the formation of the United Kingdom. Scottish entrepreneurs, excluded from English West Indian colonies, financed an ill-fated attempt to establish the Scottish Darien colony on an isthmus in Panama in 1695. About a quarter of Scotland's wealth was lost in the venture. Its failure led to the inevitable conclusion, union with England.

The dark side of sugar cane exploitation was the intensive labor it required. One estimate is that for every two tons of sugar consumed the life of one black slave was consumed2. Arabs were the first to import large numbers of African slaves to cultivated cane fields in the region of Basra. The Zanj slaves from East Africa revolted in 869 C.E.

Planting and harvesting took place at least twice a year. The working day was from 6am to 6pm with a break at midday. The ground was cleared and in each metre square a segment of cane was planted in a hole 20cm deep. At harvest the cane was cut stripped of leaves and tied in bundles to be carried to the mill. At the mill the canes are chopped up and crushed between giant rollers to squeeze out the sweet sap. The liquid which contains fats, wax and gums is purified at first by boiling. On the plantation it was passed between five or six boilers where it was boiled to concentrate it. The scum on top was ladled off. Finally, the sugar solution was allowed to cool and the sugar crystallize.

The native Carib people had been decimated by disease and genocide and could not provide the labor. Initially the English colonies used labor imported from the homeland. Freely indentured labor was augmented by convicted Royalist rebels and other convicts but these were always in short supply and soon the English had adopted the Spanish practice of importing slaves from West Africa. By the mid-18th century England with its naval superiority was dominating the slave trade. About a third of the merchant fleet was engaged in it. The Middle Passage from W Africa was to become notorious for its bestial treatment of its human cargo. About 1/8 of all slaves died on the passage. About 9-11 million slaves survived but their future was scarcely better. Their life expectancy was only 29. Their nutrition was very poor based corn flour with a little animal protein salt meat or fish, though in some cases they were allowed to grow their own vegetables, yams, cassava, sweet potatoes; plantain, banana, breadfruit, and legumes . If they resorted to chewing the young canes or tried to escape appalling punishments were inflicted on them. The value of a slave was low, about half a ton of sugar in 1700, and one ton might represent the life-time production of a slave. But the mark-up was very great. A slave bought for £3.00 in Africa could sell for £25 in the West Indies. Many died on the passage. Wastage was not compensated by the birth of children and the average price of a male slave rose from £20 in 1709 to £50 in 1780. Slaves were traded for raw sugar and rum which was, after the Return Passage traded in Bristol and Liverpool. The Outward Passage, the third leg back to Africa was with firearms, cloth, salt, trinkets. On each leg of the triangular trade profits could be made.

So the sugar trade drove the development of the modern world economy based on trade and overseas investment. It also played its part in the burgeoning of the industrial revolution. In the late 18th century a quarter of a million workers in Britain supplied the sugar/slave trade. The elegant houses of Bristol and Bath were built on sugar fortunes as the sugar millionaires bought themselves into the English aristocracy. There were at one time 120 sugar refining factories in England with an output of 30,000 tons of table sugar each year. Crystalline sugar had ceased to be a scarce luxury.

The process of sugar refining used today is quite complex, in which some stages mirror the age-old processes. They differ slightly for canes and beets. Canes are first washed, sliced up into a pulp and then pressed. The sweet juice which is expressed is then clarified in a process called carbonation; calcium hydroxide solution and carbon dioxide are bubbled through so that chalk-lime forms gathering up impurities in the process. The lime mud containing the impurities falls to the bottom and is separated from the purer liquid. The liquid is then concentrated by boiling at a low temperature to prevent caramelization under a negative pressure, to produce a clear brown syrup. Sugar crystals are then formed by seeding the syrup with pulverized sugar and the crystals are centrifuged off to produce a golden brown crystalline sugar. Molasses or dark treacle, used in the preparation of some rich foods and in some animal feeds, still bears the taste and smell of the sugarcane, a slightly bitter-salty flavor underneath the sweetness. The pure crystals are covered by a thin film of molasses containing the remaining plant impurities, so they are washed in warm pure "affination" syrup to produce a batter-like "magma" which is centrifuged again to remove the crystals. These are then washed and dissolved to about 50% strength and the clear golden syrup decolorized by passing it through an activated carbon filter or ion-exchange resin. The clear syrup is then evaporated under negative pressure and seeded with crushed sugar again to produce white crystalline sugar that is centrifuged off again. Pure sucrose crystals form preferentially and are removed.

The presence of other sugars, especially glucose and fructose inhibits this process. Alternatively a clear golden syrup that doesn't crystallize is produced from sucrose by inverting it, by the hydrolysis of sucrose to glucose and fructose, effectively producing the same sugar constitution as honey, by subjecting a sucrose solution to acid and heat. Invert sugar is sweeter than sucrose and so is used to replace part of the sucrose in many sweet drinks, thereby reducing their calorific content but maintaining the same level of sweetness. Invert sugar also has a greater affinity for water and so is used in confectionary in order to keep them moist longer. The pure sucrose crystals are dried and passed over screens to separate out crystals of different sizes suitable for castor (or caster) sugar, named for the fine sugar shaker, or caster, the sugar is fine enough to pass through, or table sugar. Icing or confectioner's sugar is mechanically crushed finer and has a little starch added to stop it from clumping.

The search for an alternative source of sugar was already decades old when the British naval blockade disrupted supplies of cane sugar to the European continent in the Napoleonic era. red. Beet had been grown for its leaves and as an animal feed in Roman times but it was not until 1747 that the presence of 4% sucrose in the root was reported by the Prussian chemist Andreas Marggraf. Many years later his pupil Achard selected a high yielding variety called "White Silesian Beet" with 6% sucrose and developed a process for sucrose extraction. Sugar beets are washed and sliced into thin strips called cossettes which are placed in a diffuser so that hot water can be pushed up through them absorbing the sugar. With sponsorship from the King of Prussia beet-sugar refining was established by 1802 in Kunern. In 1811 Napoleon established schools for the study of beet and required it to be grown. Over subsequent decades sugar beet breeding raised the sugar content to over 20% but ever since its production has been supported by subsidies and import restrictions on cane sugar.

Other plants also provide syrup, and not just in their fruits. Sweet sorghums (sorgo) for example, produce sweet pith and are cultivated for syrup. Jaggery is also a strong sweet liquid is also made by evaporating palm sap to concentrate the sugars. The manna that fell from heaven and fed the Israelites while they were wandering in the desert has been suggested is the sugary sap exuded by several different species damaged by insect attack such as *Tamarix mannifera* or *Hammada salicornia*.

Sugar, either from cane or beet, holds a central place in the western food economy. The whiteness, the purity of sugar is misleading. Sugar is a preservative, and not just in jams. Sugar and syrup are base ingredients in almost all processed foods. It is the leading culprit the world's obesity crisis. Children are addicted to sweets and sugary soft drinks. And sugar has a hidden demon; within each sugar molecule lies the alcohol molecule and it's released by the process of fermentation.

Alcohol

The demon alcohol has its hands on all sorts of levers in the brain by attaching to various receptors for neuromodulators that are secreted into the extracellular space and neurotransmitters that are secreted into the synapse between neurons, and also altering the activity of enzymes. It generally has a sedative effect, eventually leading to sleepiness, but before that it reduces inhibitions and anxiety. At each stage in a night out in the pub it works its effects.

Alcohol is a small molecule, soluble in water and lipid that directly enters the bloodstream, in the stomach and mainly the small intestine. It also easily crosses the blood-brain barrier . The brain responds in particular to the changing levels of alcohol in the bloodstream. There are several ways the brain responds. Firstly by decreased neural transmission in GABA systems . Gamma amino butyric acid or GABA is receptors are widely distributed in the brain, especially in the cortex, and are associated generally with motor ability and vision but also with relaxation or anxiety. Alcohol binds with GABA receptors in the brain and allows chloride ions to enter the neurons, thereby reducing their excitability, and leading to relaxation.

Secondly alcohol also leads to increased opioidergic (opiate producing) activity producing endorphins. Endorphins are named because they are "endogenous morphines". They act as strong pain-killers but also induce euphoric feelings. Besides behaving as a pain regulator, endorphins are also thought to be connected to physiological processes including euphoric feelings, appetite modulation, and the release of sex hormones. Alcohol results in the increased secretion of one of the most powerful natural opiates, beta endorphin, from the hypothalamus. There are several classes of endorphin receptors   Different opiate receptor types have been labeled as analgesic, δ emotional, λ sedative and σ psychotomimetic . Beta endorphin attaches to μ -receptors.

In turn increasing opioidergic activity facilitates dopamine release. Dopamanine is an inhibitory neuromodulator and neurotransmitter and has a central role in the "reward system" in the brain that is associated with feeling pleasure but it is also associated with addictive behavior. There are five types of dopamine receptors labeled D1 to D5. D1 associated with cocaine induced euphoria. D3 associated with dependency. Upregulation of D2 receptors sustains alcoholic consumption; one drink is just not enough and then there is one for the road .

Three parts of the brain are particularly associated with the reward circuit; the ventral tegmental area, the *nucleus accumbens* and the prefrontal cortex. The ventral tegmental area is located in the midbrain, and contains the dopaminergic neurons that innervate the nucleus accumbens, the prefrontal cortex, which has a role in the processes of attention and motivation, and other areas such as the amygdala and hypothalamus in the limbic system that mediates our emotions. The dopaminergic pathway that connects the ventral tegmental area and the limbic system is called the mesolimbic pathway. Another dopaminergic pathway, called the tuberoinfundibular pathway is of interest because it connects the hypothalamus to the pituitary where the greatest concentration of endorphins are found. In particular alcohol increases the concentration of beta endorphin in the ventral tegmental area. The beta endorphin interacts with mu-opioid receptors here so inhibiting GABAergic transmission. Dopaminergic neurons have GABA receptors and as a result of the lower concentration of GABA they show an increased secretion of dopamine especially in the region of the *nucleus accumbens* that acts at the interface of many different components of the reward circuit.

Thus in various ways an increasing alcohol concentration in the brain induces a feeling of euphoria and relaxes the drinker, removing his or her inhibitions and increasing their self-confidence. Unfortunately this is not backed up by increased physical ability, rather the opposite. Coordination suffers, and drinks are knocked over as the drinker stumbles to the toilet. Worse yet it explains the karaoke singers who think that they sound like Frank Sinatra, or don't care if they don't. As the drinker chatters away he or she is getting less and less likely to remember anything because alcohol also reduces the excitatory effect of glutamate on the NMDA receptors that mediate the ability to learn and remember. The drinker continues to feel thirsty because the acetylcholine receptors are also being interfered with, and now because the drinker is feeling more and more amorous and the receptors for acetylcholine are also responsible in the central nervous system for some primary emotions like anger, aggression and sexuality, the likelihood of a fight outside the pub at the end of the evening is getting more and more likely. It can go the other way, though, acetylcholine receptors are also associated with wakefulness and attentiveness, so the drinker falls asleep in his seat. It is time to stagger home to bed, he is not cold and when he stumbles and gashes his knee he feels little pain because of the increased beta endorphin.

Primitive hunter-gatherers may have been unable to make alcoholic drinks. Ethnographic studies of surviving hunter-gathers have indicated that they may not have been physiologically adapted to it . But this may only be another aspect of the wishful thinking, as if they were innocents in the garden of Eden. There was one source of alcohol that was almost ready made. This is Riddle 25 of the Exeter Book – a book in the library of Exeter Cathedral dating from the second half of the 10th century.

"I am man's treasure, taken from the woods,
Cliff-sides, hill-slopes, valleys, downs;
By day wings bear me in the buzzing air,
Slip me under a sheltering roof-sweet craft.
Soon a man bears me to a tub. Bathed,
I am binder and scourge of men, bring down
The young, ravage the old, sap strength.
Soon he discovers who wrestles with me
My fierce body-rush-I roll fools
Flush on the ground. Robbed of strength,
Reckless of speech, a man knows no power
Over hands, feet, mind. Who am I who bind
Men on middle-earth, blinding with rage?
Fools know my dark power by daylight."

The answer to the riddle is mead, probably the oldest alcoholic drink, it's such an easy enough thing to make. The honey has the glucose and all that is necessary is to allow honey water to ferment in the absence of air. Yeasts that carry out the process are everywhere and would not have to be introduced artificially.

American Indians made what they called balché flavored with the leaves of *Lonchocarpus* , a legume genus better known as a source fish-poisons and insecticides. A rock-painting of a shamanistic figure from the Tassili-n-Ajer plateau of Algeria drawn when the Sahara still had flowing rivers has a bees face hinting at the use of mead is ecstatic ritual. The Ancient Egyptians, Assyrians and Greeks all made it. The Romans drank spiced mead called metheglin. Mead survived the longest as an important alcoholic drink in northern latitudes because in the south wine made from grapes became the alcoholic drink of choice. The Celts apparently made mead with added hazel tree sap.

Mead was not an everyday drink; honey was too precious a commodity For the Slavs and Scandinavians mead was a drink of power, the drink of warriors. The best the lower ranks could get was small mead: after the honey comb had been squeezed to get the honey out it was boiled in water to free the wax, and the remaining weak sugar solution was then allowed to ferment. High mead made for the Lord's table was made directly from honey. Mead was the drink of celebrations like marriage. There is a rather unlikely tale that says origin of the word honeymoon is said to be derived from the Scandinavian tradition of the husband abducting his wife and drinking a glass of mead each day for the period of a month while he kept her in hiding. It is more likely to refer to the sweetness of early married bliss that fades with the waning moon!

The vine

"Kee yayin y'samach l'vav enosh,"
"wine cheers the hearts of men." Psalm 104:15.

In the Epic of Gilgamesh Enkidu is seduced by a temple harlot.

"Enkidu, eat bread, it is the staff of life; drink wine, it is the custom of the land.' So he ate till he was full and drank strong wine, seven goblets. He became merry, his heart exulted and his face shone. He rubbed down the matte hair of his body and anointed himself with oil. Enkidu had become a man."

Enkidu has become a man by eating bread and drinking wine, celebrating the origin of civilization, and also making the link between wine and sex.

The grape is rich in glucose. Wine fermented from grape juice early became the alcoholic drink of preference in most warm regions. It remains at the heart of any special meal. As Plato said

"Nothing more excellent or valuable than wine has ever been granted by the Gods to man."

However like all plant products used for pleasure its history is also ambivalent. Dionysus, originally the god of mead, became the god of wine. Originating as a lesser Mycenaean god, he was elevated to cult status about 540 B.C.E. He was called Bacchus by the Romans. Dionysus was the son of Zeus and a mortal woman. He taught the art of vine cultivation and gave the gift of wine, but he has two natures, bringing joy, health and divine ecstasy or brutality and unthinking rage. He was accompanied by satyrs, spirits of wild places, Silenus a drunken obese man, Pan the half-goat shepherd and fierce Centaurs, and by the Maenads, drunken women bearing rods tipped by pine cones (a reference, perhaps, to the use of resin to preserve wine) who might go mad and rip apart and eat animals raw. Dionysus was associated with rebirth after death like the vine growing back after it is pruned. In later Roman times the symbol of the liknon, a phallus rising from a winnowing basket full of grapes and other fruits is associated with Bacchus. Wine conferred a feeling of power as if by drinking it the drinker gained part of the divinity of Dionysus himself.

Wine was also associated with creativity. Most of the Greek plays were first performed at the feast of Dionysus. Nietzsche contrasted the creativism of Apollo with that of Dionysus: the first cool, structured, full of meaning and controlled; the second unpredictable, instinctual, wild, ecstatic, pleasurable and emerging from uncontrolled Nature. The Bacchae are the female followers of Dionysus, who dance in a frenzy to celebrate his rite. In Euripides play of the same name, Pentheus the teenage king is ripped apart by the Maenads who include in their number his mother. Red wine is color of blood and there are echoes here with the use of sacrifice at festivals in Lesbos and Chios, later replaced by flagellation. At the festival of Agrionia in Boetia a young boy was immolated. There are echoes even in the use of wine in the eucharist as a celebration of the Christ's sacrifice and rebirth, and of the Christian faith. In the first century B.C.E. the Roman authorities tried to suppress the Bacchanalian revels, out of concern for their potentially subversive nature. However the orgiastic revelry continued to attract the hedonistic.

The vine genus *Vitis* is a widespread genus of 60 or more species that grow throughout the northern hemisphere. Wine may have first been made from wild vines in the Caucasus but it has been domesticated in several places. It is strange that despite the availability of suitable grapes in N. America wine was not made there but utilized a variety of other plants. Grapes (*Vitis vinifera* and other species) are widespread throughout the northern hemisphere. When the Asian grape was introduced to North America it hybridized to native grape species especially *V. labrusca*. Grapes (*Vitis vinifera* and other species) are widespread throughout the northern hemisphere. The grape vine was probably domesticated in

South West Asia Domestication and viticulture followed perhaps as early as 6000 B.C.E. Sumerians may have imported some wine, from the Zagros mountains of western Iran but here palm wine was probably more important. The center of the later Assyrian empire was further west closer to areas more suitable viticulture. Wine was well-known to the Babylonians and Ancient Egyptians but beer was the most important drink, and wine reserved for important occasions with religious or symbolic significance, and then drunk until the participant in a banquet was intoxicated. *Plus ça change, plus the même chose!* According to Herodotus the Babylonians imported wine from Armenia in palm-wood casks, in boats filled with straw that floated downstream . In the Mediterranean basin, in Egypt, by the third millennium B.C.E., several different wines were being produced, the most famous being from Lake Mareotis and Tanis. The climate was probably somewhat cooler and wetter than at present. Tomb paintings provide details of all stages of the process. Viticulture spread to reach Greece about 3,000 years ago. By the time of Herodotus time most of the wine was being imported into Egypt in earthenware jars from Greece and Phoenicia.

Wherever it was grown wild grapes had the potential to cross with domesticated varieties, thereby introducing locally adapted genes. French grape varieties for example are close to the wild grape *V. vinifera* ssp. *sylvestris* from France and Tunisia. In each place grape varieties diversified. Sixteen wine famous grape varieties from northeastern France, including 'Chardonnay', 'Gamay noir', 'Aligote', and 'Melon', have a DNA fingerprint that indicates a shared ancestry with from a single pair of parents, 'Pinot' and 'Gouais blanc', both of which were widespread in this region in the Middle Ages, though the Gouais variety is very poorly regarded today and hardly grown.

Wine was normally diluted with water before consumption but this did not necessarily mean that restraint was being practiced. Symposia were drinking parties in which wine loosened the tongue. The host determined how much the wine should be watered . In Plato's symposium well-watered wine was drunk, because the participants had binged the previous night. Nevertheless the participants were mostly drunk and incapable by the morning. In the Mediterranean region in classical times wine was not reserved for the richer classes but was provide even to slaves. Cato recommends between 7 and 10 amphorae of wine per man each year. Since amphorae contained between 17 and 27 liters this almost amounts to a 75 cl bottle of un-watered wine per day.

By the second century B.C.E. most wine was being transported in the universal storage vessel, the amphora. Amphorae from south-west Italy have been found from Britain to north-west Africa around the coasts of the western Mediterranean and along the river valleys. One estimate is that 40 million amphorae were unloaded in southern Gaul in the first century B.C.E. Wine was bartered for slaves in Gaul and Diodorus Siculus writes that:

"The Gauls are exceedingly addicted to the use of wine and fill themselves with the wine which is is brought into their country by merchants, drinking it unmixed, and since they partake of this drink without moderation by reason of their craving for it, when they are drunken they fall into a stupour or state of madness."

By the end of the century the earliest French vineyards of Narbonensis were beginning to compete with Italian imports. This was the seed of French domination of the wine trade that was to last for the next 2000 years.

In southwest Spain there is some evidence of a separate origin of viticulture by about 2000 B.C.E. However, others have dated modern wine production to 1600 BC and the Greek Islands, with the invention of sealed amphorae that allowed aged wines to be produced rather than the short-lived cloudy fermented fruit juices that had previously been available. The key feature is to keep oxygen out. The Romans started to use barrels in the 3rd century C.E. but wine kept in barrels could only last a year. Retsina is said to have originated by the practice of sealing amphorae with pine resin. Resin added to the wine itself improved the preservation of wine. Wine from Chos was heavily resinated. Various varieties were available in classical times. Sweet Pramnian wine that is mentioned by Homer, was probably made from dried grapes. A similar wine called Omphacites from Lesbos is mentioned by Diosocorides. Wine from Cos was sour. Wine was flavored by a variety of other additives such as seawater, honey and spices that either acted to preserved the wine or concealed its vinegary nature . Wine was normally watered before it was drunk. The alcohol sterilized the water and it could be drunk in large volume.

The penchant for sweet wine among the rich was perhaps the most important engine driving the development of the regular trade between northern Europe and the Mediterranean region throughout the Middle Ages, that kept its culture relatively homogeneous after the Roman era. Sack was a strong light-colored sweet wine of the sherry family that was made in the Jerez region of Spain for export (*sacar* means export). It was for the time an unusually long-lasting wine though it was only aged for a year or two. It was the wine taken by explorers like Columbus and Magellan and stolen by pirates like Sir Francis Drake from the Spanish. Sack became very popular in England fueling the Merrie England of Falstaff. After the fall of the Roman Empire the use of sealed amphorae disappeared. Vintage or mature wines did not become available again until corked bottles started to be used in the 17th century, so throughout the Middle ages wine was drunk young before it went off.

Red wines is generally produced from black grapes but in Chianti white grapes are used too. Similarly white grapes produce white wine, but in the production of Champagne black grapes are used too. The important difference between red and white wines is not the color of the grape but that for white wine is that the skins are removed before fermentation. It is color a flavors, especially from the skins, pips and stalks that give red wine its color and extra flavor, but tannins are also derived from the wood of the wine barrels. As the fermentation progresses a cap of detritus from the pressing such as the skins and pips arises. The separation of the must from this detritus produces a wine that is designed to be drunk young because it does not significantly mature with age.

Three kinds of changes are involved in fermentation: yeasts convert sugars to ethanol (alcohol); acetic acid bacteria convert ethanol to ethanoic acid (acetic acid); and lactic acid bacteria convert malic acid to lactic acid. The first creates the wine must and the latter two are likely to spoil the wine must. A number of different yeasts are involved. For example, in the fermentation of red Bordeaux wine must, the fermentation is started by *Kloeckera apiculata* and as the alcohol concentration builds up Saccharomyces cerevisiae, with other more ethanol-tolerant species of yeast becoming more important as fermentation gets close to completion.

It is in the cap that the presence of oxygen allows acetic acid bacteria produce acetic acid. This process is limited by keeping the cap moist by breaking it up or spraying it with must (remontage). The addition of sulfur dioxide at the beginning of fermentation inhibits the activity of acetic acid bacteria and wild yeasts, and the exclusion of air from vats and barrels prevents spoilage. The life-time of wine was extended by the introduction of the use sulfites (Campden tablets) over 300 years ago to sterilize equipment and also added to the must. The sulfites have a dual action killing bacteria that might spoil the wine but also scavenging oxygen that allows spoiling micro-organisms to oxidize the wine.

Soft wines like Beaujolais designed for early drinking are vinified under carbon dioxide pressure, a process called macération carbonique. The use of sulfur and the importance of the exclusion of air were known to the Romans but this knowledge was lost. The lactic acid bacteria require less oxygen for their activity and although they can add an unpleasant smell they also reduce its sourness and add some more carbon dioxide, a slight sparkle in the bottle in wines like Vinho Verde that are traditionally made with higher acidity grapes.

In high latitudes the grapes are rather low in sugar and extra sugar is added before fermentation, though this process, which is open to abuse, has a posh name "chaptalization" from its inventor Chantal. When fermentation is complete a new vin de goute has been produced that is run off leaving a sediment of skins and stalks that are pressed to release a dark highly tannic wine called vin de presse. Blending of these in different quantities creates different kinds of wine. The pressed sediment can then be distilled to make a coarse spirit called marc (French), grappa (Italian) or bagaçeira (Portuguese). Secondary fermentation in the bottle is encouraged in sparkling white wines, most notably Champagne, by the addition to each bottle of a little sugared wine and yeast. Fortified wines like Port arc made by the addition of extra alcohol, in brandy, before fermentation has completed, stopping the fermentation early and maintaining a degree of sweetness that depends upon when the alcohol is added.

Wines are aged in bottles. By aging, or élevage, a wine can develop its full character dependent upon the grape variety used. The range of grape varieties provides wines to be savored of such subtle difference and specialty. Grapes with a low tannin content like Merlot do not age as significantly as those with a high tannin content like Cabernet Sauvignon. Wines designed to be drunk quickly are fined immediately, cleared by the addition of coagulants like egg-white, gelatin or isinglass and/or filtered. Vintage wines are first allowed to mature for several years, two years for Bordeaux, in wooden casks and then fined before bottling. Racking, transferring wine from one cask to another leaves the sediment behind and introduces small quantities of air so that the malic acid fermentation can take place. After bottling tannins and anthocyanins condense and sediment and rich esters are formed.

Famous varieties of vine

Cabernet Sauvignon - Deep red, Complex,depth, fruit flavors, blackcurrant, blackberry, long maturing

Merlot   Red, More acid than above and faster maturing

Shiraz, Syrah   - Red, Luscious, silky, spicy

Grenache (Garnacha)   Red or Rosé   Sweet, fruity, low tannin

Pinot Noir -   Red, Delicate fruits and flowers to rotting vegetables, perhaps the oldest variety

Gamay - Red , Beaujolais, acid but low in tannin

Chardonnay – White, Light and subtle but often heavily oaked

Sémillon -White, Dry or sweet, yellow, with hints of citrus

Sauvignon Blanc – White, Light with a touch of dryness, enjoyed young or slightly aged, made more robust by fermenting in oak

Riesling - White, German, Sweet fragrant with a touch of spice

Muscadet - White , Extremely dry, and light

Muscat - Red or White, Provides fruit as well as wine

In the 15th century in Germany it was realized that wine in large barrels aged better than that in small barrels; we now know because this limits exposure to the air and the activity of acetic acid bacteria. Some huge barrels were constructed; culminating in a tun constructed in Heidelberg in 1663 held 37,500 gallons. However by this time another technological solution was underway; wine in bottles sealed by cork stoppers. Bottling was first used widely in the production of the new high quality wines of the 17th and 18th century notably Bordeaux and Champagne. The fermentation of wines around Epernay and Reims was stopped by the winter chill before it was complete and started again in the bottle in the Spring. These fizzy champagne wines first became very popular in the late

17th century in Restoration England. By this time English glassmakers were making strong bottles to contain the pressurized wine.

Meanwhile vines were being introduced into the new European colonies with mixed success. The Spanish introduced the cultivation vines to Mexico and thence rapidly throughout Latin America and later to California. The Dutch introduced vines to the Cape and the British introduced vines to Australia and New Zealand. Notably the introduction of European vines to Virginia was a flop. The European vines succumbed to American diseases. In the 19th century vines imported back to Europe from North America led at the introduction of a powdery mildew called *Oïdium* probably on specimen plants grown in glasshouses. By 1851 the vineyards of Europe were suffering an epidemic of mildew. Quickly a means of controlling it by sulfur treatment was discovered but not before resistant North American vines had been widely introduced. On some of them was a much greater threat the *Phylloxera* aphid. Its effects were first noticed in the 1860s in the Côtes du Rhône but in twenty years it had spread throughout France and on to Spain, Italy and Germany. Eventually resistant American vines were introduced as root stocks for the European varieties, though local resistance to the American vines, now seen as the source of the problem, the Franco-Prussian War and the expense of uprooting whole vineyards delayed the final triumph over Phylloxera until the 1890s.

These two crises encouraged the scientific study of viticulture and vinification, and a shift to larger scale production. Louis Pasteur first showed that yeast was responsible for fermentation. Scientific progress has not halted and led in the late 20th century to the new wines, of consistent quality produced in the Americas, South Africa and Australia. The techniques imported back into Europe greatly improved the quality of European wines from the periphery that had long been regarded as second class. The 2000 year hegemony of France was at last challenged. Now red wine is promoted as valuable for health as moderate red wine intake may protect against cancers and heart are also because it is rich in anti-oxidant compounds, flavonoids, especially the phenolic compound resveratrol. Disease, also found in dried fruit raisins, sultanas (seedless) and currants (from Corinth)

Cider and other alcoholic beverages

In more northern climates it was not the grape but other fruits, especially the apple, that provided sweetness and from which the demon alcohol was released. However the origin of domesticated apples and perhaps even cider making was central Asia. Particular apple varieties noted for their sweetness that were propagated by grafting, a technique known to the Romans. Cato noted several varieties of apple, and Pliny the Elder 22 varieties. Today there are thousands of named varieties. Apples *Malus domestica* have a self-incompatibility that makes sure the outcross so that each seed is genetically unique. Occasionally a natural crossing with wild crab apples M. sylvestris or *M. pumila* would give rise to a new variety.

The whole process was repeated in the white-man's colonization of America. The colonists took grafted apples with them but these did not thrive. However crossing with American crab apple gave rise to new varieties suited to the American climate. However even hard and sour apples had an important place, perhaps the most important place, because they could be used to make cider. The orchards planted from apple seed were for cider not for fruit. The trees were pippins, literally grown from apple pips. Ironically Johnny Appleseed's and others' planting of seeds for cider apples was a glorious breeding experiment – because a few trees grew up that were sweet including Golden Delicious and Jonathan.

Cider making is as ancient as any other fermentation. Its origin may be the mountains of Kazakhstan, the center of origin for domesticated apples. The Hebrews made Shekar and the Greeks Sikera by boiling apples with fermented juice. The Basques call apple trees Sagara from the same root word. Cider and Cidre (French) are from the same root. It is supposed that the Druids introduced scrumpy making to the British Isles. Scrumpy is hard cider, made entirely from apples fermented naturally. Perry is made from pears in the same way. Wild yeasts bring about the fermentation. The art of cider making probably spread west with the Celts. The Greek Stabon recorded the abundance of apples in Gaul. By the 9th century C.E. Charlemagne was ordering Sicetores to brew cider and perry on his estates. Traditionally farm workers were paid partly in cider several pints per day, more during harvesting and haymaking!

Cider making was well-established in the Basque country well established before a cider making industry was established in the Cotentin district of Normandy in the 11th and 12th century CE . A cider industry was established in England in the 13th century. It was a drink of the poor competing with ale and beer but not wine. Somehow cider avoided the moral doubt that wine provoked in those with a puritanical bent but applejack, a strong alcoholic beverage made by distilling cider, 30-40% alcohol, did not. One way of making applejack was to freeze cider; the water freezes first and could be strained out.

Unfortunately most ciders, the ones sold as sweet or medium cider, are made from concentrated apple-juice, sugar and water.

Freeing the demon

The brewing of beer and the baking of bread are partners . Indeed the production of bread may be an outcome of brewing . The oldest surviving recipe from 3,800 years ago is part of a hymn to Ninkasi the Sumerian goddess of brewing. In Ancient Egypt a common greeting was bread and beer and sent their children to school with both. Almost every crop has been utilized to produce alcohol. All starchy crops can be used if at first the starch is first converted to sugar and the glucose released. The process is as varied as the kinds of drinks created. For example in East Africa beer (busaa in Kenya, merissa, an alcoholic porridge, in Sudan) is made from a variety of grains such as maize, sorghum or millet or starchy root crops such as cassava, generally using malted millet or maize to break down the starch or sugars to release the glucose for alcoholic fermentation.

Malted finger millet has a higher amylase activity than either sorghum or maize, and equivalent to barley, and hence mobilizes sugars more effectively. Malted sorghum has another disadvantage in that it contains high quantities of dhurrin, a cyanogenic glycoside that is hydrolyzed to cyanide HCN (also called prussic acid or hydrocyanic acid) when the rootlets and shoot are damaged. Removal of these reduces the cyanide content of the malted grain by 90% . Another problem are the high tannin content of the grains. High tannin varieties are grown because they discourage bird predation. The sorghum is first soaked overnight in a slurry of wood ash in water; the alkaline conditions reduce the tannin content. The grain is then drained and allowed to germinate before being sun-dried. Pounding loosens any ash that adheres and the sprouts that are high in cyanide. The grain is then ground and used to prepare either a non-alcoholic beverage called obushara or an alcoholic drink containing about 3 % alcohol called omuramba.

Another source of alcohol is the fermented sap of the cocoanut palm *Cocos nucifera* (called mnazi in Kenya) to produce toddy a kind of palm wine. Palm wines are made from a variety of different species of palms around the world, especially in the lands surrounding the Indian Ocean. After flowering the tip of the palm inflorescence is cut and allowed to drip for seven days. Then the sap is collected and allowed to ferment naturally, by the action of wild yeasts, to make toddy or palm wine. Within 24 hours an alcoholic beverage is produced with 6% alcohol. The sap can also be allowed to evaporate to make treacle or jaggery. Arrack is distilled from the fermented toddy. Other palm genera used in this way include *Phoenix*, *Borassus*, *Raphia*, and *Nipa*. Astonishing amounts of sap can be obtained; from a single *Caryota urens* (kithul or toddy palm) more than 20 liters of fermentable sap can be obtained, about 10 liters per inflorescence.

Two kinds of enzymes mediate the metabolism of alcohol in humans; alcohol dehydrogenase and aldehyde dehydrogenase. The first converts alcohol to aldehyde which is toxic, but this is quickly converted into acetic acid by the second. Each enzyme is coded for by a family of genes and there are several different variants (alleles) present in different proportions in different human populations. There is clinical evidence that some of this in the C.E.H enzyme is related to the high rates of alcohol dependence in some human populations like some American Indian populations, but also to the tendency to binge drink or even the flushing response to alcohol . Individuals who flush when they have an alcoholic drink have an, accelerated heart rate causing an elevated blood flow and dizziness, sweating and nausea, but this discourages them from binge drinking. The C.E.H alleles associated with flushing, and are in high frequency in Asian and Jewish populations . ALDH I deficiency has a frequency of 25-50% in East Asia and the indigenous population of South America.

Alcohol has also a long history of medicinal use. The strong beer produced in classical times called *Zythum* was a component of many medicinal remedies. Medicines were dissolved in it and in previous centuries it was given in great quantities to the ill. At least if drunk in large enough quantity it kept the patient quiet. It was not until the late Victorian era that hospitals were founded in Britain on the basis of strict abstinence. But it is not so long since Guinness was marketed as being "good for you". Although today advertisers might be wary of claiming medicinal value for alcoholic beverages who can doubt the profound pleasure and soothing of the soul that alcohol can bring and not just from intoxication.

By volume the most important alcoholic drinks are made from grain. Barley is the grain used most often to make ale or beer and spirits. It is first soaked and then allowed to germinate to produce the malt. The enzyme α-amylase converts the starch to sugar in the germinating seed. The germinating grain is turned to ensure it is aerated then germination is brought to a halt by baking so that the mobilized sugars aren't wasted by the growth of the seedling. Brewing was a process established by the Sumerians. First they germinated emmer wheat and barley seeds and then coarsely ground them to make a flour that they baked into small cakes. The cakes were crumbled and soaked in large earthenware pots to ferment to create an alcoholic beverage called *Sikaru*. The key discovery of the Sumerians, associated later by the Babylonians with the god Sirus, was how to release the sugars for fermentation by first combining barley (normally rich in α-amylase) and wheat though wheat varieties also rich in α-amylase are also know. The Egyptians followed the Sumerian/Babylonian method of malting and produced beers with increasing strength called busa, zythum and dizythum. Zythum was flavored with juniper, ginger and saffron. In some ways the Russians brew kvass, produced by adding pieces of stale black rye bread to rye malt and allowing the mixture to ferment is an echo of this ancient method of brewing. Wheat, barley, and buckwheat meal are also used but an important addition is of sugar, either as fruit, birch sap, or as refined sugar. Apple or raspberry are favored fruits and the addition of raisins maintain the fermentation even after bottling, though kvass normally has only a low alcoholic content (0.7 to 2.2%).

The process used in the industrial breweries today is to grind the malted grain in a mill to make grist. The grist is then mixed with water to make porridge-like mash. Then the mash is left to allow the dissolved sugars to leach out forming a thick liquor called wort that can be drawn off and the remaining mash washed with hot liquor to remove the final sugar as that. Alternatively the mash is drawn off little by little and boiled up in a cooker to encourage the sugars to diffuse out, and returned to the mash tun, before the wort is finally drained away when it is strong enough.

The wort is boiled in a brew-kettle or copper, normally with hops, and then cooled. The temperature and length of time baking helps to determine the flavor. Hops is now normally added to beer for increased bitterness and aroma but originally because it acted as a preservative. The hop *Humulus lupulus* is a climber related to Cannabis sativa. It also prolongs the life of beer. Beer is hopped ale. The use of hops in beer is first recorded in the Finnish saga Kalevala, dated by some to 1000 B.C.E. but the first reliable record dates from the 9th century . By the 12th century its use was well-established in Germany, Bavaria is still the world's largest hop growing area, though even in Germany there was fierce competition from traditional ale-brewers at first. The Archbishop of Cologne, who had a monopoly on herbs used to flavor ale, tried to suppress its use. Later Dutch ale-brewers complained about beer from Hamburg but then started brewing their own beer. The cultivation of hops spread west to the Netherlands and then, in the early the fifteenth century, into England by Flemish cultivators, though hops is native to England. In the Tudor period the English brewers of un-hopped ale complained about the competition gaining the support of King Henry VIII who banned the royal brewer from using it . Nevertheless by the early 17th century the brewing of ale was dying out as beer brewing spread .

Hop is a dioecious vine and in most areas it is cultivated as a seedless female, propagated asexually. The female inflorescence is made up of a cone-like collection of bracts which have at their base glands that produce a yellow resin called lupulin the bitter agent. Hops aren't the only source of bitterness in beer. At different times tannins from oak and ash trees were used in Scandinavia, cinnamon in southern Europe and sassafras, liquorice and sweet fennel in America. The hop vine which is supported on wires supported by poles many meters high dies back to the ground in winter. There are two basic early types of hops, the Hallertauer sort from Bavaria, and the Saaz type from the Czech Republic. The Golding type hop has been cultivated in England for about 250 years. It has a flowery bouquet and is used for dry-hopping, the addition of dry hops to the beer in the cask to add extra aroma. The Fuggle variety arose by chance in 1861 and was cultivated by a Kent hop-man called Richard Fuggle. It quickly rose to dominate the English acreage. American varieties arose by hybridization between introduced European varieties and the native American hop. In recent years high-alpha varieties with more bitterness have been developed., and with resistance to disease. Some older varieties have almost been lost; Hallertauer

Lambic beers brewed in the region around Brussels and to the southeast perhaps use the most ancient of all methods. A mash of 40% un-malted wheat and 60% malted barley is boiled with old hops and then left to cool in shallow open vats at the top of the brewery where open louvers allow wild yeasts to float in. Once fermentation has started the beer is allowed to ferment in wooden casks for up to three years! Lambic beer is rather tart or sour to the taste and may be sweetened with caramel or used as the basis of a cherry (Kriek) or raspberry (Framboise) beer. "White" beers like Hoegaarden are cloudy wheat beers flavored with coriander and curaçao. Another flavored beer is Sahti from Finland flavored with juniper. The flavoring of ales with other ingredients is an ancient practice. Honey and dates were used to flavor Sikaru and the Ancient Egyptians added emmer wheat not hops for flavor and also used juniper, ginger and saffron. The ales of Turkey from 700 BC may have been flavored with honey and fruit or mixed with mead and wine! Ale was so highly thought of by the Sumerians that it was used to honor the gods and treat sick people. The ale was drunk from a communal vessel using reeds as straws to avoid the debris. to their own strong beer made called heget or in Greek zythum. Like sikaru it was made from fermented bread but this was squeezed and filtered out. A stronger ale called dizythum and a weaker ale called busa were also brewed. Ale making in Egypt declined with the spread of Islam.

Lager is brewed with the yeast *Saccharomyces carlsbergensis* that settles on the bottom giving a clear beer. Pilsner is a lager brewed with hard water (higher in magnesium and calcium). Ale is brewed with Saccharomyces cerevisiae that floats. It makes a darker stronger beer than lager. Porter and later stout are brewed with roasted barley and/or roasted malt, resulting in a darker, more strongly tasting and potentially more alcoholic beer.

Like the vine, barley as the source of alcohol has obtained a mythological life. Even more than wheat, barley is the staff of life. The myth of John Barleycorn as in the version of Robert Burns' makes explicit the link between the preparation of barley to the Christian symbolism of death and rebirth.

There was three kings into the east,
  Three kings both great and high,
And they hae sworn a solemn oath
  John Barleycorn should die.
They took a plough and plough'd him down,
  Put clods upon his head,
And they hae sworn a solemn oath
  John Barleycorn was dead.
But the cheerful Spring came kindly on,
  And show'rs began to fall;
John Barleycorn got up again,
  And sore surpris'd them all.
The sultry suns of Summer came,
  And he grew thick and strong,
His head weel arm'd wi' pointed spears,
  That no one should him wrong.
The sober Autumn enter'd mild,
  When he grew wan and pale;
His bending joints and drooping head
  Show'd he began to fail.
His coulour sicken'd more and more,
  He faded into age;
And then his enemies began
  To show their deadly rage.
They've taen a weapon, long and sharp,
  And cut him by the knee;
Then ty'd him fast upon a cart,
  Like a rogue for forgerie.
They laid him down upon his back,
  And cudgell'd him full sore;
They hung him up before the storm,
  And turn'd him o'er and o'er.

They filled up a darksome pit
  With water to the brim,
They heaved in John Barleycorn,
  There let him sink or swim.
They laid him out upon the floor,
  To work him farther woe,
And still, as signs of life appear'd,
  They toss'd him to and fro.
They wasted, o'er a scorching flame,
  The marrow of his bones;
But a Miller us'd him worst of all,
  For he crush'd him between two stones.
And they hae taen his very heart's blood,
  And drank it round and round;
And still the more and more they drank,
  Their joy did more abound.

Brewing has always been an integral part of celebration and ritual. There is the tradition of holding a wake.

Until the late Middle Ages in England brewing was done in the church leading to a great piss-up on the village green and no-doubt in the cemetery. One wonders how many new lives were started by drunken revelers using the gravestones to indulge in a bit of rumpy-pumpy – new life and death brought together by alcohol.

Anytime, anywhere ....anything

Apart from malted barley and wheat a wide range of other starchy plants are utilized to brew ales and beers, and a variety of methods are used to release the sugars from the starch. Tesgüino was made from sprouted maize in America. Chicha is a South American beer made from maize or quinoa. Chicha de Jora is made from malted red corn "Jora", Chichi Blanca is a white corn version with added cinnamon, Chichi Picante has added lemon and aji peppers. Chicha de Maní is made from quinoa and peanuts. Traditionally a chichi jar is put on the roof for good luck. Traditionally chichi was made by chewing the grains and spitting them into a vat for fermentation. There is a large quantity of amylase ptyalin in saliva that breaks starch down to the glucose (and maltose) required for fermentation, the first step in releasing the alcohol molecule within.

Traditionally a similar technique was used to make sake from rice, a technique that dates from China 4000 B.C.E. The rice, and also nuts, was first "polished" in the mouths of villagers squatting around a communal tub, into which they spat the mush. Later it was discovered that milling and the introduction of Koji would do the same job and allowed mass production. Koji-kin or koji mould is a mould called *Aspergillus oryzae*, a mould similar to that in blue cheese. It is cultivated on steamed rice to allow it to make the enzymes that break the rice starch down to sugars. Following the introduction of mass production in the 1300s Sake became the national Japanese drink.

Starchy root crops were also used. Cauim is the traditional alcoholic beverage of Brazil from pre-Columbian times. It uses a process similar to that traditionally employed to make chichi. Although maize grain is sometimes also used the main ingredient is cassava. Thin slices of cassava are boiled to make them tender then allowed to cool. Then they are chewed, it is women's' work, and spat into a pot for further cooking and then the paste is allowed to ferment in a large pot. The cauim is used in large quantities at celebrations of different sorts, warmed first and served by the women to drunken dancing men. Today cassava is harvested in Brazil for industrial alcohol production. Parakari is fermented by Amerindians in Guyana from cassava using the amylolytic mould *Rhizopus* that frees the sugars from starch. Mobbie was made in the West Indies from sweet potatoes.

Bottling the Genie

It was through the Arabs that distillation entered Europe. It was one of the primary skills of the alchemists and practiced by alchemists in the region of Salerno by the 11th century. The words alcohol and alembic (meaning a still) are derived from the Arabic words al-koh'l and al-anbĭc though the latter has a Greek root from the word for cup or beaker. The work of Arabic alchemists became more widespread in Europe through Raymond Lull (1233-1315) from Palma in Mallorca who traveled widely in the Middle East and North Africa and Arnaldus de Villanova (c1235-1313) who taught botany and alchemy at Barcelona, Montpellier and Paris.

Aqua vitae, distilled wine or brandy, with an alcoholic strength 8-9 times greater than wine, was at first mainly used in medicine but by the 15th century it had become a popular drink, so much so that in 1496 the city authorities in Nuremberg forbade its sale on feast days. In France Louis XII took its production away from apothecaries and alchemists and granted it to the guild of vinegar makers . The strong wine survived transportation better and had a longer lifetime than ordinary wine and so became popular in the northern parts of Europe, especially in the Netherlands where it gained its name brandewijn or burnt wine. The making of brandy became common in those French wine regions used to exporting to the Netherlands especially in the Charente region, but also elsewhere, wherever it was a useful way of turning low quality wine into a more saleable product. Inevitably there was also a premium product produced from the Columbard vine around Cognac.

The fashion for brandy was greatly helped by its supposed medicinal properties.

"It is good for them that have the falling sickns if they drink it. It cureth the palsy if they be anoynted therwith... It sharpeneth the wit, it restoret memory. It maketh men mery & preserveth youth..... If gargild it remedieth the diseas in the throte.... it expelleth poison. "

Rather than "*restoret memory*" if it didn't cure you it made you forget your ailments: another lines reads
"It is merveylous profitable for frentik men & such as be melancholy".

Its use medicinally is ancient. For example its right use is given in the Caraka Samhita a Sanskrit text from the first centuries of the Common Era. Alcohol was one of the drugs of Arabic medicine . John Brown in the 18th century, saw all illnesses as being a disturbance of function, requiring treatment either by opiates for their sedative effect of alcohol for its stimulant effect. One of the main treatments provided in hospitals was the administration of alcohol. In 1873 the London Temperance Hospital was founded but even here the use of alcohol was not outlawed but only discouraged.

The technology of distillation destroyed the wine monopoly of highly alcoholic beverages. Highly alcoholic drinks could be distilled from grain alcohols. Whisky is normally distilled from a mash only of barley. The different flavors come from the different qualities of the water, yeast strains and the casks it is matured in. Bourbon is distilled from a mash of grain containing, according to US federal law, not less than 51 percent corn, along with barley and either wheat or rye. Each distillery has its own unique blend of grain and some of the mash recipes are generations old-family formulas jealously guarded. In East Africa busaa can be distilled to make a spirit called chang'aa that is up to 60% alcohol.

Some notable spirits have other sources of carbohydrate than grain. Vodka is made from grain or from potatoes. Tequila and Mescal are made from Agave. They are usually clear in color and unaged, distilled from the fermented juice of the Mexican plant, specifically several varieties of *Agave tequilana*. Mescal, is a similar beverage to Tequila, is less expensive and stronger in flavor, and is made from an agave plant that grows wild in the Oaxaca region. Pulque from the maguey. Rum and Cachaça are distilled from sugar cane juice allowed to ferment for a very short time, in some for about 24 hours only. Rumbullion or Killjoy was first produced in large quantities in the mid-17th century in the West Indies. In a short while it was widely traded in the American colonies, England and the Netherlands. It was even traded for slaves in West Africa. In Brazil sugar cane sugar is used to ferment industrial-style alcohol aguardientes and also a premium white rum called cachaça that is aged in oak barrels. It is used to make caipirinha, a lime and sugar cocktail, and batidas with fruit juice, milk and ice.

Distilled alcohols and their botanicals

Distilled alcohol is a pretty flavorless substance. Other plants add flavor whether it is the oak barrels used to age whisky or the wood charcoals used to filter bourbon. Distilled alcohol was used as the basis of many fruit based liqueurs, aperitifs or ratafias. Rosoglio or rossoli was made from sweet raisins was made in Italy in the Middle Ages. Benedictine was first created in 1510 at the Abbey of Frécamp in Normandy by a monk called Bernardo Vincelli from distilled wine, honey and 27 different botanicals. The passion for experimentation with different herbs went to ridiculous lengths. Parisian monks created Chartreuse with 130 botanicals.

Gin gets its name for Dutch word *genever*, meaning juniper. Gin is distilled from grain, not necessarily entirely barley. Genever was originally a medicine and indeed its reputation as a means of bringing on a miscarriage may be related to the effects of juniper. But it was originally distilled by Franciscus Sylvius de la Boe in the University of Leiden because he thought that the juniper berries would provide a remedy for the tropical fever that plagued the Dutch in the East Indies. Gin is distilled from grain, not necessarily entirely barley, and flavored by "botanicals", such as juniper, seeds, coriander seeds, orange and lemon peel, angelica and orris root, and cardamom pods. Bombay sapphire gin has 10 different "botanicals" the most important being juniper berries and coriander.

As well as adding flavor botanicals have also been added to magnify the effect, to add bite. The Dionysian cult magnified the access to the natural world by use of fennel *Foeniculum vulgare* (Apiaceae) mixed in wine. Alcohol has always been a vehicle for conveying medicaments anyway as well as being regarded as a medicine in itself. Absinthe as a drink made with *Artemisia absinthum* arose from this medicinal use. Perhaps surprisingly it has a fairly recent origin. It was little more than a century, between the writing of the original recipe written by Dr. Pierre Ordinaire in 1792, and the founding by Henri-Louis Pernod of the most important absinthe distillery in France in the early 1800s, to its banning in France in 1915 because of is dangers. In the meantime it became favored in the bohemian circles of France for its ability to stimulate creative activity. Its users and abusers included the poet Rimbaud, the writer Baudelaire and the painter Van Gogh among many others. Absinthe is an emerald green alcohol, due to the presence of chlorophyll, with added herbs (including aniseed, fennel, hyssop, lemon balm, angelica, star anise, dittany, juniper, nutmeg, and veronica), the most important being the bitter wormwood *(A. absinthium)*. The name wormwood denotes its former use to counteract parasitic worms. Vermouth is made from the flower heads of wormwood gets its name from the German for wormwood (wermuth). Up to 90% of wormwood oil is Thujone, also isolated from *Thuja occidentalis* and other plants. The psychoactive role of thujone is uncertain but the related species *Artemisia nilagirica* was smoked in West Bengal for its psychoactive effects, and Artemisia caruthii was inhaled by the Zuni native Americans as an analgesic. Calamus (*Acorus calamus*) and nutmeg (*Myristica fragrans*) were also sometimes used in making absinthe and may have enhanced the psychoactive effect. Absinthe was diluted with cold water poured over a perforated spoonful of sugar turning the shot of absinthe milky white as the essential oils were precipitated out of the alcohol. Much of the effect of absinthe was purely alcoholic. Undiluted absinthe had 60-85% alcohol so a large part of its peril was associated with alcohol abuse.

Wormwood is used to flavor the Swedish brannvin made from potatoes. Chartreuse made by Carthusian monks supposedly following an ancient recipe called the "Elixir of Life" contains 130 herbs and spices, contains small amounts of thujone. So does Benedictine made by Benedictine monks. Other drinks no longer use wormwood. Herb Sainte, from New Orleans, and Pernod are wormwood-free absinthes but contain Star-anise (*Illicium verum*) for flavor. Herb Sainte is manufactured in New Orleans. Pastis is a similar liqueur to absinthe and was also originally made with wormwood, but is now flavored with liquorice (*Glycyrrhiza glabra*). Other essential oils are used top flavor many other drinks like Ouzo and Jägermeister.

The Alcohol Industry

Brewing started out as a home and kitchen activity. Over time it is an activity that has become more and more centralized and controlled. Governments of all Ages have interfered and sought to control the production of alcohol.

The home and kitchen activity became replaced by specialist brewers, a few of which became some of the largest and most important industries in many towns, but there were very many of them. With improved transport there was a tendency for larger firms, the ones that could invest in a large scale production could gain an extra margin and so the larger breweries grew at the expense of the smaller ones. However the live nature of ale and beer meant that it didn't travel well, so that even by 1900 there were 6,000 breweries in the U.K.

A century later there were only 500 and the market is dominated by just a few. By the 1980's the six largest breweries owned more than the pubs and brewed 75% of the beer. Worldwide the brewing industry exhibits in a marked degree the growing control of our food and drink by fewer and fewer suppliers. Trade in alcohol has always been an important activity and the trade routes established, and in previous times, for example, in Europe between the southern vineyards and the north provided also a route for the flow of ideas between northern and southern Europe, in effect binding the continent together in a shared culture. Now similar brands are available worldwide.

The scale of the brewing industry is staggering. Beer still forms he largest part of the market in value and volume, then sprits and before wine. The top 6 beer manufacturers control more than 50% of the worldwide market, and taking alcoholic beverages as a whole only three companies control 40% of the market.

In the U.K. the breweries economized on production and sold their beers and lagers, nitro-keg or pasteurized versions, that traveled well, across the whole country. Concerns about the lack of competition led to the 1989 Beer orders that prevented any brewery owning more than 2,000 pubs but instead of giving up their pubs the major breweries created separate pub companies to which they sold their pubs. The new *PubCos* marketed beer in much the same way. Landlords rented their pub from the PubCo and also have to buy all their beer from them, generally at above market price, the price the PubCos pay to the brewer. One result is that the major breweries supply 80% of all beer and lager drunk. And the homogenization of the pub experience began in earnest as pubs were modernized to a standard company theme.

In more recent years there has been a resurgence of "craft" brewing and even the production of craft spirits. The rise of the micro-brewery is based on technological advances that allow small batches of beer to be produced regularly to a high quality. The major brewers and distillers have noted the trend and have jumped on the wagon, taking over micro-breweries or establishing their own craft ranges. "Craft" is well on its way to becoming just another word to use in marketing.

The social effect of alcohol consumption

Alcohol, along with their imported diseases, played an important role in the Europeans gaining ascendancy over the aboriginal Americans. As early as 1637 Thomas Morton a settler who preferred the company of the Massachusetts Indians to his compatriots wrote

"Although drunkenness be justly termed a vice which the savages are ignorant of yet the benefit is very great that comes to the planters by the sale of strong liquor to the savages who are much taken with the delight of it .... I shall have no trade, if I will not supply them with lusty liquors; it is the life of the trade. ".

Later in the French and Indian Wars of the 18th century the French were at first able to maintain a large alliance of the Iroquois against the Anglo-Americans because they were seen as the source of supplies amongst which wine and brandy were well to the fore. Later after the defeat of the Indians across North America and the destruction of their culture alcoholism became one symptom of their degradation.

The industrialization of the alcohol industry wider availability of highly alcoholic drinks fueled an explosion in alcoholism and alcoholic hedonism from the middle of the 17th century onwards. In England the restoration of the monarchy signaled a break out from the strictures of Puritanism of the Commonwealth.

Alcohol has fueled the work of many artists and writers and destroyed the lives of not a few. Unfortunately the excessive use of alcohol does not guarantee the production of great artistic work or else we could all be proper artists and not just piss artists.

Alcoholics and sugarholics

Alcohol is metabolized primarily in the liver where it is first converted to toxic acetaldehyde by the enzyme alcohol dehydrogenase. Acetaldehyde has a variety of physiological consequences, such as flushing of the face and neck, nausea, headache nausea and drowsiness, but is normally rapidly metabolized to acetate by aldehyde dehydrogenase. Variations in the forms of these enzymes have a geographical pattern and are to some extent correlated with the different ability of people to tolerate alcohol. Many Asians such as the Chinese and Japanese have two gene variants, especially one of the aldehyde dehydrogenase gene, that protect them from alcoholism because they slow the conversion of acetaldehyde to alcohol. They are protected from alcoholism because drinking alcohol is not a pleasurable experience. These gene variants are not present in Europeans, native Americans, Indo-Malayans, Pacific Islanders or Koreans who do not suffer the symptoms that make the Chinese and Japanese averse to alcohol. Koreans are 50 times more likely to become alcoholics than the Chinese.

However alcoholism and other alcohol related behavior are complex traits. The effect of alcohol depends upon not just alcohol metabolism but also rate of absorption and transport. The contrasting behavior of the French who consume their alcohol with a meal and the binge drinking British is only one example. Food in the stomach greatly delays the absorption of the alcohol. After consuming the same amount of alcohol over the same period the Brit is drunk and the Frenchman still *compos mentis*. The Frenchman is just as likely to suffer from liver damage though.

Several different genetically controlled traits, including alcohol dependence, the level of response to alcohol, the presence of coexisting depression, or the maximum number of drinks a person consumes per occasion, which have a strong genetic component, influenced by genes on several different chromosomes. Aldehyde may be related to averse behaviors in the periphery but in the brain may be reinforcing. Several other enzymes, apart from alcohol dehydrogenase, such as cytochrome P4502E1 and catalase contribute to aldehyde production.

Alcohol leads to the release of beta endorphin and dopamine. Eating high carbohydrate foods is linked to the release of serotonin, beta endorphin and dopamine in the brain. There is a link between obesity and alcoholism in that some people have a greater risk of become either an alcoholic or obese because they have a lowered level of serotonin functioning and an augmented response to beta-endorphin within the reward systems of the brain , . It has been proposed that people who have a sweet-tooth, so-called sugar sensitives or chocaholics, have naturally lower levels of brain serotonins and dopamines and so are more sensitive to the foods which lead to their release . Individual differences in drinking behavior and risk of alcoholism may be related to differences in D2 dopamine receptors in the nucleus accumbens consumption . Interestingly in experiments alcoholic men show a preference for high sucrose drinks compared to normal people . One wonders if this may lie behind the preference of many alcoholics for strong sweet lagers or for spirits mixed with a sugary mixer like rum and coke. Spirits mixed with coca cola also have the extra stimulation of the caffeine they contain.

Chocolate, of course contains molecules that potentially might have a direct effect on the brain, caffeine, but also theobromine though these are in low concentration. Several other chemical have been discovered in the more than 300 different chemicals that have been identified. These include phenylethylamine, a natural amphetamine, and even anandamide, a neurotransmitter related to THC the active chemical from Cannabis. However these chemicals are in so low concentration that it is unlikely that they are responsible for chocolate craving. Chocoholics do not generally show a chemical dependency, they are just addicted to the lovely taste, the sweetness, and the mouth feel of chocolate that causes all their endogenous pleasure transmitters to be released.

Unfortunately in obese people, and alcoholics, the brain has adapted to the higher levels of dopamine that food and alcohol induce and they have lower numbers of dopamine receptors than normal. Now a high carbohydrate or alcohol intake is required to maintain the same amount of pleasure as normal people.

6 Secrets of the Ancients

The earliest written record of the medicinal use of plants is a Sumerian Herbal written by Enlil-bani. Clay tablets from 2100 B.C.E. mentions myrrh, cypress, opium poppies, caraway and thyme.

Imhotep is usually represented as the "Mummy" who comes to life an wreaks havoc, his name means "he who came in peace" and as well as being vizier, the first of the master architect of the step-pyramid, he was an Egyptian physician. He lived around 2600 B.C.E. and founded a medical tradition that was highly influential. He is thought to be the original author of the Edwin Smith Papyrus, a surgical text that makes little reference to herbal remedies. The earliest herbal dates from 1500 B.C.E. and is a 20m long papyrus purchased in 1874 by the Egyptologist Georg Ebers from Edwin Smith. On it there are listed 876 prescriptions made from 500 different plants. The Edwin Smith Papyrus is remarkably straightforward but the links of herbalism to mysticism and religion were well-established. In the Ebers papyrus there is the invocation "Come Remedy! Come thou who expels evil things in this my stomach and limbs."

The knowledge of plants gave power and was jealously protected - different kinds of plants were given secret sacred names. *Ambrosia maritima*, a ragweed, now used to flavor liqueurs, was called vulture's heart. The knowledge of the use of herbal medicines has been cloaked in mumbo-jumbo ever since. Theophrastus (c. 372-287 BC) a student of Aristotle and later became his friend and collaborator, the father of a scientific study of plants, makes fun of the superstitions of some herb gatherers who advocated, for example, the collection of peony roots at night lest their eyes were pecked out by woodpeckers. Another superstition was that the only way to pull mandrake from the ground was to tie a dog to it and from a distance call the dog with a horn. Pulled from the ground the shriek of the mandrake killed the dog but out of earshot the herbalist was spared. It's a scene that is played out again in Harry Potter.

Knowledge of plants was a dark art, the secret skill of witches. Henbane *Hyoscyamus niger* and Deadly Nightshade *Atropa belladonna* were particularly associated with sorcery. Their use is rooted in myth and legend. The dead in Hades were crowned with henbane as they wandered hopelessly beside the Styx. Symptoms of poisoning are dilated pupils, dry skin, thirst and hallucinations leading to a deep "drunken" sleep. The same symptoms were evidence of the prophetic vision smoking these plants induced. The oracle at Delphi may have sat over a crack in the rock whose trace inducing emanations were empowered by the addition of smoldering henbane. Their tropane alkaloids hyoscyamine and hyoscine (sometimes called scopolamine) are found in 22 genera of the Solanaceae including *Datura*, the thorn apple. The plants with the highest levels of tropane alkaloids are two species of corkwood from Australia *Duboisia*. The range of uses and reported effects of these plants is very wide but not unusual. Eyedrops of *Atropa* dilated the pupils to make them more beautiful – hence its name "belladonna" – beautiful woman. Henbane was reputed to be aphrodisiacal too, stops diarrhea, and it was also used as a pain-killer, an analgesic, for example to counteract toothache in ancient Egypt. Like some other plant drugs there is a therapeutic, a toxic and a psychoactive effect (more about the last of these in the next chapter).

Poisons

Some of the earliest references to the power of plants over human physiology are about their use as poisons. Socrates, convicted of not recognizing the gods of the state, introducing new divine things, and corrupting the youth was sentenced to drink an infusion hemlock (*Conium maculatum*), which unusually for the umbels (Apiaceae) is rich in alkaloids. On the island of Chios a form of euthanasia was practiced using Aconite, monk's hood or wolfs bane *Aconitum napellus* (Ranunculaceae). It contains a range of toxic alkaloids aconitine, lappa-aconitine and batrachotoxin which can be absorbed through the skin It was famous as particularly effective against women. Nicander (d. 130 B.C.E) wrote that if it touched the female genitals death occurred within 24 four hours. One Roman Calpurnius Bestia was reputed by Pliny to have eliminated four of his wives in a row. Rhazes tells of a girl who ate so much aconite that she built up a resistance to the toxins but her kisses poisoned her lovers and even her chickens avoided the places where she spat. More prosaically picking the flowers may be dangerous: only 1 g of crude plant parts can cause death .

Poisons have also been used to kill people in many parts of the world as part of the judicial process as well as criminally. The Calabar Bean *Physostigma venenosum* was used in a trial by poison by the Efik tribe in Nigeria. One bean was enough to kill a person but if the accused vomited up the poisonous potion they were innocent. But such powerful drugs have also been used as medicines. Today *Physostigma* is the source of physostigmine that is used in the treatment of glaucoma. *Gelsemium elegans* (allspice jasmine) is well known to the hill tribes of South East Asia as a means of committing suicide .

Hundreds of different species are listed in databases of poisonous plants. In different parts of the world different plants have provided arrow or fish poisons. Most are sources of alkaloids. A different species of *Stephania* (*S. hernadifolia*) is used as a fish poison in Australia. Aconitine was used in various parts of the northern hemisphere as an arrow poison. *Strophanthus* and *Strychnos* in Africa and Asia and the latter also in South America, where other sources, *Chondrodendron* and *Curarea* (both Menispermaceae) and *Hura* (Euphorbiaceae) provided other poisons of remarkable efficacy. *Hura* provides a poison half a million times more toxic than potassium cyanide.

Many poisonous plants have more often been used to save life rather than take it. An example of a class of compounds that are potentially toxic are the cardiac glycosides found in plants like the foxglove Digitalis and Pheasant's Eye Adonis. The arrow poison strophanthine from *Strophanthus* is one. It is said to have been discovered when one colonial botanist stored his toothbrush with some poison arrows and found after brushing his teeth that his heart beat strongly. Cardiac glycosides act on the heart to increase the force and speed of contraction so that the heart has a longer rest period between contractions but if in a toxic dose they can cause arrhythmia and heart failure.

Foxglove was traditionally used as laxative and its use as a heart tonic was not established until the 19th century. Digoxin, derived from lantoside C, from *Digitalis lanata* is now a widely used drug to stabilize and strengthen heart function. Dr. William Withering establishment of the use of Digitalis to treat dropsy, is regarded by some historians marking the beginning of the scientific method in medicine. He noticed that a patient suffering from dropsy recovered after a local folk cure. The cure contained twenty different herbs but Withering thought that one, foxglove, used traditionally as a diuretic and purging agent was probably the active one. He also knew it was potentially fatally poisonous. However Withering was an outstanding product of the 18th century British enlightenment. As a pharmacists son he had risen to become chief physician at Birmingham General Hospital. He became a member of the "Lunar Society", so named because they met on nights of the full moon, a society that included such luminaries as Joseph Priestley, Matthew Boulton and James Watt; it was a society founded on the ides of enlightenment ideas of rational explanation and scientific experiment. After a series of careful experiments, using no less than 156 different subjects, he established a safe dose of foxglove. One wonders what happened to those subjects he gave too high a dose to. In 1785 he published *"An Account Of The Foxglove And Some of Its Medical Uses: With Practical Remarks On Dropsy And Other Diseases"* perhaps the first modern work in pharmacology.

Adonis vernalis has the glycosides cymarin and adonitoxin among others. Its name is rooted in Assyrian and Greek myth as a god of nature. *Convallaria majalis*, Lily of the Valley, with cardiac glycosides such as convallarin, was used as an anti-dote to poison gas as well as a heart tonic. A problem with cardiac glycosides is that they have a low therapeutic index; meaning that the a dose that is toxic is not much greater than one that has a therapeutic value. One leaf of oleander *Nerium oleander* contains enough of the cardiac glycoside oleandrin to so disturb the regular heart beat that cardiac arrest results. In one incident in the invasion of Spain by troops of Napoleon, native people forced to provide meat to the invaders, mixed with it oleander, killing many of the French troops.

The dark art of making poison potions and their passed down through history. Italian princes were reputed to be obsessed with growing plants that might be antidotes. Pietro D'Abano's *"De Venenibus"* written in 1300 became the standard text book on poisoning. Lucretia Borgia was reputed to have a hollow ring in which she hid poisons. But the understanding of poisons was widespread. The foliage of yew *Taxus baccata* was well known to have no antidote. Shakespeare has Macbeth prepare a poisonous brew, which included "slips of yew" and some authorities think that the "hebon" poison poured into the ear of Hamlet's father may have also been composed of yew; another name for yew is "hebenon". Yew was a sacred tree, associated with death and rebirth. It is green in mid-winter and roots readily from the branches. Bulls sacrificed to Hecate in ancient Rome were wreathed in yew foliage. And yet like many other poisonous plants it also has therapeutic value as the source of an anti-tumor drug (see below). The power of plants is amply represented by yew, a dark mysterious power represented in myth, that can confer life or death.

Dioscorides and after

Modern herbalism in Europe can be dated to the Ancient Greeks. Theophrastus seems to have respected, and quotes from, the Athenian herbalist Diocles who had collected together the existing knowledge of medicinal herbs. Only fragments of Diocles written work survives, but it probably provided the foundations of all later works by herbalists like Crateus, doctor of King Mithradates of Pontus, and Sextus Niger each of whom who made their own original additions. The culmination was the herbal produced by Dioscorides in about 60 C.E.

The "*Materia Medica*" was the foundation of botanical knowledge for a millennium and a half even though its descriptions of plants were poor. Widely read, and widely travelled, probably while serving as a doctor in the armies of the Emperor Nero, Dioscorides was able to incorporate his own good sense and experience in the work. It included nearly 600 plants. Luckily a beautifully illustrated version of the "*Materia Medica*" has survived called the "*Codex Vindobonensis*". Produced in 512 C.E. for Juliana Anicia the daughter of Flavius Anicius Olybrius the Emperor in the West, the codex has a chequered history. It first turns up in 1406 in a monastery in Constantinople. After 1453 with the conquest of the city, it was in the hands of the Turks. The Jewish doctor of Suleiman the Magnificent seems to have purloined it. Busbecq, that adventurous ambassador of the Holy Roman Emperor, who will return to our story later, saw it and managed to get some drawings for Mattioli. Seven years later a sale had been negotiated and the Codex arrived in the Imperial library in Vienna.

The importance of the Codex is that there is good reason to believe that many of the drawings, which are very naturalistic in style, are derived from earlier ones drawn from nature, perhaps even by Crateus himself. The *Codex Vindobensis* represents a peak of botanical knowledge and observation. For the millennium after it was produced there was a sad decline in the quality of copies of Herbals and scarcely any new observations were made.

The collapse of the Roman Empire almost extinguished botanical knowledge in the west. Only Christian monks kept a flicker of the classical expertise alive. One remarkable exception from the generally poor quality of medieval Herbals was the *"De Vegetabilibus"* of Albertus. He was ironically nicknamed Albertus the Great because he was very short. The Pope once asked him to stand up and not weary his knees in an audience until he realized that Albertus was already standing! As a Dominican friar and Papal envoy Albertus travelled widely across Europe botanizing on his way. Once he was sent to Poland to stop the Poles killing and, sometimes eating their unwanted children. In parts of *De Vegetabilibus*, which was written between 1250 and 1260 Albertus records his own acute observations of plant structure. This marks him out from all other herbalists and encyclopedists of the day, who merely repeated or collected others work in a more garbled form.

The generally low standard is illustrated by *"De Proprietatibus Rerum"* produced by Bartholomew at about the same time as Albertus was writing. It became the standard authority on natural history until the 16th century but confines itself to restating traditional accounts many derived ultimately, via many other authors, from Dioscorides and Aristotle. He shows no knowledge of garden flowers, except for rose, violet and lily, but is more interested in fables like the one about hunted Cretan wild goats eating Dittany (*Origanum dictamnus*) to eject arrows from their wounds.

The Doctrine of Signatures

Herbal medicine is as old as humankind though at times it has been plagued by superstitious nonsense. As early as the Classical period Dioscorides derided the beliefs of some kinds of folk doctors. Plants with red organs, like the red seeds of peony were prescribed for menstrual problems or bleeding. The dark-purple loosestrife, *Lysimachia atropurpurea*, was used to stop bleeding so-called *machia* from *lysimachein* = "causing strife to cease". The hardness of the seeds of gromwell, *Lithospermum officinale*, from *lithos* = stone and *spermum* = seed, meant it should be used to break up kidney stones. The quackery present in many works had a long history. Pliny had reported the use of the ashes of a rose gall, looking like a ball of down, to be mixed with honey and applied to a bald head, to make hair grow. Man's vanity and credulity has not changed much in two thousand years. Some remedies by happy accident were effective. The root of mandrake, *Mandragora officinalis*, shaped like a man, was especially powerful. It was used to put people to sleep before surgery, and used along with henbane and opium, but if its strength was miscalculated the anesthesia became permanent!

This kind of herbal symbolism was formalized an irascible medic called Paracelsus. By his doctrine of signatures the shape of herbs indicated the remedy they provided. This theory arose from the medieval attitude that all nature was created for mankind. The doctrine of signatures had a resurgence in the 17th century promoted by writers like Robert Turner who stated in *"Botanologia, the British Physician"* published in 1664,

"God hath imprinted upon plants, herbs, flowers as it were hieroglyphics the very signature of the vertues" .

Aristolochia clematitis - birthwort, with a curved flower like the curved womb and fetus was used as a remedy in difficult pregnancies, *Aristolochia* meant birth-improver. Walnut with its seed shaped like the brain was good for mental ailments. Plants with red organs, like the red seeds of peony were prescribed for menstrual problems or bleeding. The dark-purple loosestrife, *Lysimachia atropurpurea*, was used to stop bleeding . The hardness of the seeds of gromwell, *Lithospermum officinale*, meant it should be used to break up kidney stones.

The quackery present in many works had a long history. Pliny had reported the use of the ashes of a rose gall, looking like a ball of down, to be mixed with honey and applied to a bald head, to make hair grow. Man's vanity and credulity has not changed much in two thousand years. Some remedies by happy accident were effective. The root of mandrake, *Mandragora officinalis*, shaped like a man, was especially powerful. It was used to put people to sleep before surgery, and used along with henbane and opium, but if its strength was miscalculated the anesthesia became permanent!

The Conundrum of Homoeopathy

The doctrine of signatures seems risible today and yet many people swear by homoeopathy which is as inexplicable in terms of scientific understanding. Samuel Hahnemann founded homoeopathy with an article published in 1796 and in "*Organon der Heilkunst*" of 1810 which espoused what are now called the" classical" homoeopathic principles of "like cures like", the "single remedy", the "minimal dose" and the "potentized remedy". Homoeopathic principles like the minimal dose have a close similarity to some aspects of Ayurvedic medicine.

The "like cures like" principle seeks to promote the natural responses of the body to a disease by treating with a remedy that provokes a similar syndrome, so onion (*Allium cepa*) is prescribed for hay fever because it provokes the same symptoms of crying and irritation around the eyes and nose. Ipecacuanha (*Cephaelis ipecacuanha*) is used as an expectorant and causes symptoms mimicking asthma. For each disease a single remedy may be prescribed and sometimes it seems a single treatment is enough to provoke the body's own ability to cure itself into action. A potentized remedy is made from a mother tincture by dilution and shaking ("succusssion"). The mother tincture, in the case of herbal treatments, is made by steeping plant material in 20% alcohol water solution to extract its therapeutic "virtues". A remedy that is diluted 1:10 is called a 1X dilution, another 1:10 of the first 1X is called 2X and so on . A remedy diluted 1:99 is called a 1C dilution and so on. A typical product is 30C, one-part mother tincture to 100 to the power 30 (100^{30}). In order to ensure that there is at least one molecule of the mother tincture left would require the patient to drink more water than is contained in all the oceans . However "succussion" is an important part of the dilution process because by vigorous shaking of the solution the memory of the active component is transferred to the water molecules. Jacques Benveniste, a French Scientist claimed in 1988 to have scientific proof of this memory and later even proposed that the memory was electro-magnetic and could be extracted and transferred electronically. The homoeopathic paradox is that as successive dilutions are made and the original component disappears, so that not even a single molecule may remain, the homoeopathic strength of the solution increases. To most scientists and medical practitioners these practices seem hopelessly naive and lack any basis in empirical science.

Nevertheless there have been a series of experiments that purport to demonstrate the reality of homoeopathy. For example the pharmacologist Nieber at Leipzig University has apparently demonstrated potentization with belladonna in an experiment using rat intestines. However many positive results have been difficult to repeat and others have failed to stand up to intense scrutiny about the kind of statistics reported , though it is fair to say that many mainstream scientific experiments are not subject to this intensity of scrutiny. Nevertheless, balanced against the relatively few experiments that have demonstrated a positive effect, normally an extremely marginal one, are the many experiments that have shown no effect. Unfortunately homoeopathy also seems to be plagued by charlatans and obscured by mumbo jumbo. The mother tincture is made by steeping the plant, there is no control on how strong it might be, or what it might contain. There is even a rather carefree attitude in some cases about which species of a plant genus that is used. A casual look at homoeopathic web sites make it clear that many homoeopaths can't even spell, and following Hahnemann, they prefer to use names for species and other bases of remedies that make the remedy seem more mysterious. "*Natrum mur*" is table salt sodium chloride! All this nonsense clouds the fact that in many cases homoeopathy seems to work, even perhaps if precisely because of the mumbo-jumbo they are cloaked in, they trick to body into healing itself – like the well-known placebo effect.

A greater irony is that many of the plants utilized in homoeopathy do have a pharmacological effect, if they were present in any quantity. Many are used in traditional herbal medicine. A homoeopathic kit might include *Aconitum napellus* (monkshood), *Allium cepa* (red onion), *Arnica montana* (leopard's bane), *Atropa belladonna* (deadly nightshade), *Bryonia alba* (white bryonia), *Matricaria chamomilla* (chamomile), *Gelsemium sempervirens* (yellow jasmine), *Hypericum perforatum* (St. John's wort), *Strychnos ignatii* (St. Ignatius Bean), *Cephaelis ipecacuanha*

(ipecacuanha root), *Ledum palustre* (marsh tea), *Strychnos nux-vomica* (poison nut), *Pulsatilla vulgaris* (wind flower), *Rhus radicans* (poison ivy), *Ruta graveolens* (rue), and *Veratrum album* (white hellebore) all of which have a demonstrated pharmacological activity. Several are notable poisons, the ones in bold, so perhaps it's just as well that homoeopathic dilution removes all traces of them from the remedy! Of the others the evidence from clinical trials is equivocal. For example 30X Arnica showed no significant effect in reducing muscle soreness .

The Ayurveda – Life Knowledge

As ancient as the Classical herbalism of the Mediterranean region was the classical herbalism of India, the Ayurveda, from the Sanskrit *ayur* - life and *veda* - knowledge that probably dates back 5,000 years. By 800 BC different herbals mention 500 and 760 medicinal plants. Manushi is the use of plants in treatment. Herbal remedies are substances *dravyas* that work in the body by their properties *guna* such as *ushnatva* (hotness), *ruksha* (dryness) and *pichhilatva* (sliminess). Herbs are classified according to their habitat and their actions on *dosha* or body type. The medicinal properties can be increased by various treatments *sanskar*. Their purpose is to promote the body's own healing processes. In this traditional medicine the efficacy of herbs has a spiritual dimension, that is enhanced by worship and the recitation of mantras before they are collected, and as they are prepared and used.

A famous Ayurvedic herb is Kutki or Katuka (*Picrorhiza kuroa*). It stimulates the immune system, is anti-allergic and preliminary clinical trials showed an anti-asthmatic effect though a one double-blind trial didn't. It has antioxidant activity . A commercial drug called Picroliv containing the active components picroside and kutkoside has been developed.

Normally however in Ayurvedic medicine herbal medicines are used in combination. The combination of ginger (*Zingiber officinalis*), turmeric (Curcumin longa), frankincense (*Boswellia serrata*) and ashwagandha (*Withania sominiferum*) has been shown in a couple of studies to reduce swelling in rheumatoid arthritis and osteoarthritis. *Triphala* ("three fruits") is an Ayurvedic combination of amalaki or Indian gooseberry (*Phyllanthus embilici*), bibhitaki

(*Terminalia belerica*) and haritaki or herda (*Terminalia chebula*) that provides a general health tonic promoting good digestion, increasing red blood cells and removing undesirable fat. These ingredients certainly have an anti-microbial property .

Indian gooseberry (*Phyllanthus embilici*) is also the main component of *Chyavanprash* a general tonic provided as a jam that contains 40 or more different herbs. It has 30 times more vitamin C than oranges. Other components are ashwaganda, haritaki, cinnamon, shtavari (asparagus), bamboo, clove, cardamom, pippali or long pepper (*Piper longum*), guduchi (*Tinosporia cordifolia*) vasaka (*Adhatoda vasica*), Punarnava or spreading hogweed (*Boerhaavia diffusa*), musta or nut grass (Cyperus rotundus), Sati (*Hedychium spicatum*), Bhumiammalaki (*Phyllanthes amarus*), Brihati (*Solanum indicum*), Gokshuraa or caltrops or Devil's thorn (*Tribulus terrestris*), Bilva or Bael Tree (*Aegle marmelos*) and Nagkeshara (*Mesua ferrea*) (Whew!).

With this number of ingredients it seems churlish to question the efficacy of such a remedy. Indeed, these ingredients do have a variety of different activities, indeed the individual herbs have multiple components and multiple activities. *Aegle marmelos* is astringent and antiviral anthelmintic, anti-inflammatory and antimicrobial (against vibrio cholera and Salmonella) properties. In some, the active component(s) have been identified. The active component of *Tribulus terrestris* is harmine which is also a component of Ayahuasca from the vine *Banistereopteris caapi*. It blocks the action of serotonin, and is useful in kidney dysfunction and male impotence. The roots of *Boerhaavia diffusa* contain rotenoids and the boeravinones, dihydroisofurenoxanthin and boerhavine, and are reputed as anti-inflammatory, and analgesic effect. Rotenoids are also found in *Derris trifoliata* which provides the insecticidal derris dust.

Rebalancing the energy – Chinese herbalism

Chinese medicine has a proven history of efficacy, though it too is cloaked in spiritual mumbo-jumbo. It is based on rebalancing the "energy", the Qi or Chi of the body, and redirecting it to restore harmony, though what that energy might be is difficult to comprehend.

According to legend the first Chinese herbal called Pen Tsao was produced by the emperor Shen-Nung dates to about 4,700 years ago. It lists 237 herbal remedies. Another called Nei Ching Su was produced by the emperor Huang-Ti. Traditional Chinese medicine is part of an old and continuous tradition of herbalism in Asia. From the 237 herbal preparations listed in the Pen Tsao the number of prescriptions rose over the centuries to about 11,000 but in the 18th century this was cut back to use only 300 herbs of which about 150 are considered essential. Traditional Chinese medicine is becoming increasingly popular in the west. There are said to be 400-500 Chinese medicinal herbs available in the West. Medicines are available for any illness or disease.

Remedies are based on four or more herbs: the King herb has the strongest effect and is directed at the most important imbalance caused by the disease; the Minster herb is directed both to the main imbalance and secondary imbalances; an Assistant herb that either strengthens the effect of the King herb, or eliminate any harsh or toxic side effects of the King or Minster herbs; and a guide, envoy or messenger herb that directs the other herbs towards the organ or region of the body that requires treatment. Herbs have temperatures, tastes (based on the five elements - water, wood, fire, earth, metal, and directions. All this is rather fanciful and traditional Chinese medicine has also been put under the spotlight by orthodox medical practitioners and scientists and as with homoeopathy many clinical studies claiming efficacy of the remedy have been challenged . Criticisms include inappropriate experimental set-up and collecting of data, including randomization, blinding, small sample size and inappropriate controls, a focus on short-term results, a lack of quantification, lack of base-line information and a failure to report side-effects. Importantly there is a clear bias in reporting only positive results.

Many traditional Chinese herbal medicines are proving to be efficacious against a range of diseases though because they are often administered in combination with other herbs that effect has been difficult to establish. In addition commercially available herbal medicines are not produced in a standardized way and consequently have widely varying levels of any active components. The quantity of active ingredients varies from source to source and even with batches. In tests some lack any active ingredient and others have potentially dangerous concentrations. Unfortunately the quality of these "natural" products sold to the public is very variable and they are often adulterated with other components. One notable case occurred in Belgium in 1991-2. Slimming capsules that contained two Chinese herbals medicines had the climbing herb *Stephania tetrandra* with the superficially similar looking *Aristolochia fangchi*, which contains kidney toxins and carcinogens. Over 100 cases of acute renal failure were reported. Another case involved the poisoning of 51 people who drank herbal star anise tea. The tea contained *Illicium anisatum* (Shikimi fruit flowers) as well as *I. verum* (Chinese star anise) . The powdered bark and leaves of the shikimi fruit is used in incense and has insecticidal properties. In Chinese it is called Mang-thsao or "mad herb" because it causes frenzied activity in people poisoned with it.

Perhaps most herbs utilized by traditional herbal medicine will prove to have potential. About 140 medicines have already been developed from traditional Chinese remedies . Salvianolic acid B from sage species especially Danshen, *Salvia miltiorrhiza* but also other salvia species such as S. officinalis, is a rosamarinic acid dimer related to other phenylproponoids such as caffeic acid. It is a potent phenolic anti-oxidant and has proven efficacious in a variety of conditions preventing tissue damage and has an anti-dementia effect . In particular it aids blood circulation and limits the development of liver damage from chronic hepatitis or alcoholism. Clausenamide from wampi *Clausena lansium*, a distant relative of citrus fruits, is a potassium channel blocker and has potential in promoting liver recovery and as a nootropic drug (meaning acting on the mind). Traditionally it has been used medicinally for a variety of purposes to cool the stomach and aid digestion, against bronchitis, to preventing dandruff and for darkening the hair.

There have been attempts to produce standardized traditional remedies. One, for example is marketed as Equigard® It contains *Epimedium brevicornum, Morinda officinalis, Rosa laevigatae, Rubus chingii, Schisandra chinensis, Ligustrum lucidum, Cuscuta chinensis, Psoralea corylifolia* and *Astragalus membranaceus*. It is aimed at restoring the harmony of the kidney but it has been shown to act against prostate cancer cells .

The Search for a Green bullet

Traditional herbal medicine with its emphasis on the use of a combination of herbs, and treating the whole person resonates with us today because of our disaffection with modern mechanistic medicine that seems so often to focus entirely on the disease and loses sight of the patient. But modern medicine has not totally ignored our herbal tradition though until recently it has most highly valued and sought plant medicines that can act like green bullets to defeat debilitating diseases. The greatest of all these has been quinine.

Before the widespread use of quinine the one overriding danger for the explorer in the tropics was catching malaria. Malaria may have been introduced to South America by the Spanish. There are various legends and stories associated with the discovery that the bark of a tree native to South America provided a remedy. In 1633 a Jesuit priest called Father Calancha in Peru had noted its efficacy. In 1638 the wife of the Viceroy of Peru, the Count of Cinchon, lay very ill with malaria. The desperate physician successfully tried "Peruvian bark" and a remedy was borne which made the occupation of the tropics by Europeans and North Americans a real possibility. Peruvian bark was exported by the Jesuits to Rome, at that time a highly malarial location. Its use spread throughout Europe.

In the 18th century Joseph de Jussieu described the tree. Linnaeus later named it *Cinchona* after its first famous recipient. In 1820 the French chemists Pelletier and Caventou isolated the active component of the bark, the alkaloid quinine. The value of the bark was so great that the Dutch and the British attempted to break the South American monopoly. Banks was first to suggest the collection of the species of cinchona from the Andes. Clement Markham (1830-?) retired at the age of twenty-one from the Royal Navy to become an explorer. He traveled in the eastern Andes in 1852-4. Conditions were tough. In the previous thirty years there had been no fewer than 43 uprisings and revolutions. However, Markham managed to persuade the Indian Office and Hooker to fund an expedition to collect *Cinchona* trees and transport them to India for cultivation, some via Kew, some directly to Calcutta. The transfer was only made possible by the invention of the Wardian case, a kind of transportable mini-glass-house. Meanwhile in 1852 Hasskarl the

Dutch director of the botanical garden in Java entered South America under a false name and bribed an official with a bag of gold for some *Cinchona* seeds. The results of these competing efforts were a relative failure. The plants gained were low yielding varieties. An alternative source was from an Australian called Ledger who persuaded an Aymara Indian called Incra to smuggle seeds of high yielding plants out of Bolivia. Incra was successful but was later discovered and tortured to death. Ledger failed to sell his seed to Britain. They were wary because plants previously supplied by Ledger were low yielding. The Dutch government took the chance and thereby a multi-million-dollar industry was established in Java and Amsterdam based on high-yielding *C. ledgeriana*. At one time it produced 97% of the world's quinine. The South American industry based on wild trees was decimated. However the capture of Java by the Japanese and Amsterdam by the Germans in the second world gravely threatened the Allies ability to fight in the tropics. This re-ignited the South American industry and also the search for synthetic substitutes that after the war eclipsed the use of quinine.

It is shocking then to realize that malaria remains a global health problem that even surpasses the pandemic of AIDs. More than a million people, 90% in Africa, mainly pregnant women and children under five, die each year, and the problem is worsening. Paradoxically the problem has been that synthetic anti-malarials have been too successful and applied a strong selection pressure for resistance and so they lost their effectiveness. The adoption of ACT (artemisinin based combination therapy) holds much promise for a longer lasting remedy. It is interesting that in a way it follows the tradition Chinese herbalism by utilizing several drugs in concert. In a similar way the adoption of combination therapy in the treatment of AIDs mirrors the principles of traditional Chinese medicine.

An excellent example of this is the "discovery" by western drug companies of the efficacy as an anti-malarial, the ancient Chinese herbal remedy qingzhaosu, the herb *Artemisia annua* (sweet wormwood), which is a relative of the absinth . This is a very promising rediscovery because resistance to most of the synthetic anti-malarials such as chloroquine has become very widespread. The active component is a class of compounds called artemisinin, which can be administered as an herbal tea qingzhaosu, produced just by steeping the plant in boiled water for 15 minutes, but it is normally administered in artemisinin combination therapy (ACT) in combination with a companion drug such as SP (sulfadoxine /pyrimethamine), amodiaquine or mefloquine, which has the advantageous effect of both shortening the period of treatment and limiting the potential for resistance to arise. Another anti-malarial with a Chinese origin is yingzhaosu A from the ylang-ylang vine, kithali champi in Bengali, *Artabotrys uncinatus*; it is related to *Cananga* the source of the scent. *Artabotrys uncinatus* is a climber that has been used traditionally to make a stimulant tea. The drug arteflene, a synthetic peroxide has been developed from it .

Chaulmoogra oil was for a while another green bullet, providing a treatment for leprosy, against the bacterium *Mycobacterium leprae*. In the west it was for centuries thought to be untreatable. The only remedy was to isolate lepers to the fringes of society, in Scandinavia cowbells were slung round them to warn of their approach, or to confine them to leper colonies or houses. However in India ns South-East Asia Chaulmoogra Oil from the *Hydnocarpus kurzii* a tree from India, Sri-Lanka and South East Asia. In legend it was either discovered by a Burmese prince or the god Rama sent into the forest because he had leprosy. It was used for a wide range of skin conditions in Ayurvedic medicine. It was first brought to the attention of western medicine in the 1854 by Frederic Mouat the Physician of the Medical College Hospital in Calcutta. Initially the source of the oil was confused but in 1901 David Prain identified the seeds in the Calcutta bazaar as coming from *Hydnocarpus kurzii*. In southern India a different species *H. wightiana*, called Tuvakara in Sanskrit. Shortly thereafter Frederick Power at the Welcome laboratories analyzed the oil chemically and identified the active component an unsaturated fatty acid with the formula $C_{18}H_{32}O_2$ he named Chaulmoogric Acid and another called Hydnocarpus Acid.

The oil was at first applied both externally, mashed up in butter, and taken internally. The latter was more successful, though it made the sufferer nauseous, so the introduction of a method, developed in the Philippines by Victor Heiser and reported in 1913, by injecting it after it had been dissolved in camphor, was an important step in making it of more widespread use. In the following years

Chaulmoogra Oil was widely exploited but supplies were uncertain and of variable quality. This situation encouraged in the 1920s a botanist from Hawaii called Philip Rock to search for the source in India and Indo-China. Eventually he found a tree and obtained some seeds that were used to establish a plantation of nearly 3000 trees on Oahu. Today we would call this bio-piracy but the financial gain was relatively short-lived because by the 1940s alternative synthetic remedies were becoming available. The search for a synthetic medicine was encouraged by the variability of results achieved with Chaulmoogra Oil, which worked best in early stages of leprosy only. Perhaps too some of the variability experienced was due to variations in the quantities and qualities of active components from plants grown under different conditions and different seasons.

The search for further green bullets goes on. There was a time when at regular intervals a new plant was hyped as the cure for AIDS. *"Drugs only address the symptoms, herbs address the cause. Drugs simply mask the symptoms, herbs assist in the healing."*

This statement illustrates a false dichotomy that some practitioners in complementary and alternative medicine, and herbalists, have. Herbal remedies must be good and better than the drugs produced by the pharmaceutical industry because they are *"natural"* – but how does this square with the fact that plants directly provide, or have led to the development of one quarter of all prescribed medicines produced by the major drug companies? Quinine is unquestionably the plant based drug that has the greatest impact on human development, because it allowed western colonization of the tropics, is quinine but is it a drug or a herbal remedy?

Aspirin - the panacea

Some medicines have proved to be panaceas. Foremost among these and the widest used medicine of all is aspirin. It is a synthetic acetylsalicylic acid, a modified form of salicylic acid, a benzene ring with a carboxylic acid and hydroxyl group attached. First marketed by the drug firm Friedrich Bayer & Co. in 1899 as "a" *"spirin"*, was first produced by a chemist at Bayer called Hofman in 1897. The name comes from where salicylic acid had previously been obtained, from *Filipendula ulmaria* (meadowsweet), then called *Spiraea ulmaria*, and was first synthesized in 1860 by Kolbe.

Various derivatives of salicylic acid are found in plants, either as glucosides or esters. Oxidation of these in the liver and bloodstream releases salicylic acid. The efficacy of plants that produced these glycosides, salicin in *Salix* (willow) and *Populus* (poplar) bark, in *Viburnum*, populin in poplar, gaultherin in *Gaultheria* (wintergreen), and spiraein *Filipendula* has long been recognized for reducing inflammation and relieving pain. Hippocrates (460-377 B.C.E.) recommended the powdered bark of the willow and leaves against fever and Gerard (1564-1637) recommended meadowsweet boiled in wine to provide a remedy for pains of the bladder. In North American tribal peoples had independently discovered the use of willow bark against pain.

Aspirin inhibits the production of prostaglandins that magnify the pain sensation and cause swelling or cramps. Prostaglandins have many functions in the body including encouraging the sticking together of platelets in the blood making it sticky. Hence aspirin is recommended for people who are at risk from a heart attack because it thins the blood , though this anti-clotting activity can cause bleeding from the stomach in high doses, a danger that is far less significant in the natural "salicins". Aspirin also acts on the hypothalamus which controls temperature and so reduces fever. It truly is a panacea for many ills.

Plant Chemistry in Action

The chemistry of plant active compounds is extremely complex . There are about 300,000 species of plants and each of these has a wide range of active compounds. The categorization of these compounds in any simple way is impossible. For example one class are the alkaloids, alkaline organic compounds with a cyclic ring structure containing one or more nitrogen atoms, each connected to two carbon atoms. There are a huge number of different kinds, named after the plant they were discovered in, either the genus (berberine from the genus *Berberis*), or species (boldine from the species *Peumus boldo*) or their effect (emetine is an emetic) or their discoverer (L'Obel hence Lobeline). Chemically they can be broadly classified by whether they are heterocyclic or non-heterocyclic, or more narrowly as pyridine-piperidine alkaloids, quinoline alkaloids, isoquinoline alkaloids, tropane alkaloids, quinolizidine alkaloids, pyrrolizidine alkaloids, indole alkaloids, steroidal alkaloids, alkaloidal amines and purine alkaloids .

One example where the chemical structure predicts the therapeutic value are the glycosides. Glycosides are compounds that can be broken down by hydrolysis into a sugar and an active non-sugar component. Sometimes they are called pro-drugs because they are inactive until this hydrolysis occurs in the gut. Phenylpropanoid glycosides were first was discovered in 1964 in *Verbascum sinuatum* (verbascoside) and have since been identified from 60 species in 14 plant families especially advanced families in the subclass Asteridae (Scrophulariaceae *Verbascum*, *Digitalis pupurea*; Gesneriaceae *Rehmannia glutinosa* Plantaginaceae *Plantago asiatica*, *P. lanceolata*, Asteraceae *Echinacea pallida*, *E. angustifolia*, (Echinocoside – a trisaccharide like verbascoside has antibiotic and antiviral properties and proven anti-oxidants (though alkylamides and polysaccharides are of greater significance in overall action of *Echinacea*) Lamiaceae *Stachys*, *Teucrium*, *Ocimum sanctum*, and Verbenaceae *Verbena*, *Lantana camara*).

The anthraquinones glycosides, many of which act as laxatives, are present in Senna, rhubarb, aloes and buckthorn. They are yellow-brown and some are dyes, like those from *Rubia tinctoria* the source of madder and several species of the buckthorn *Rhamnus*. The most widely used plant laxative is Senna from species in the genus *Cassia* (formerly called Senna) especially *Cassia Senna*, *C. acutifolia* and *C. angustifolia*. The anthraquinone glycosides sennoside A and B, like many other glycosides, are pro-drugs and pass through the gut unaltered but are broken down in the caecum and colon by the microorganisms. There dianthrones are released which are processed further into anthrone and anthraquinone, which inhibit water and electrolyte uptake and encourage peristaltic action, thereby having a laxative effect. In a similar way anthraquinone glycosides aloin A and B obtained from juice and latex of *Aloe barbadensis*, dried and filtered, are hydrolysed in the lower bowel. Aloin-like anthraquinone components called cascarosides are obtained from the bark of *Rhamnus cathartica* common or purging buckthorn and *R. purshiana*, cascara sagrada or sacred bark. They are so powerfully laxative that they are normally aged for at least a year to allow the cascarosides to be oxidized to milder components.

A wide range of herbal medicines appear to have their beneficial effect because of their anti-oxidant properties . The anti-oxidant effect is most important in degenerative disease, preventing the signaling of damage that leads to programmed cell death (apoptosis). Many plant anti-oxidants are flavonoids and polyphenols and plants that produce higher concentrations of these have a reputation for wound healing and in anti-aging creams . They include *Echinacea*, ginseng, grape seed, green tea, lemon, lavender, rosemary, arborvitae (*Thuja*), sarsaparilla, soy, prickly pear, sagebrush, jojoba, *Aloe vera*, allantoin, feverwort, bloodroot, apache plume, and papaya. Green tea polyphenols, catechins may act to prevent programmed cell death in Parkinson's disease as well as promoting natural healing of the skin. Genestine may act to prevent apoptosis of hippocampal neuronal cells in Alzheimer's disease. Hawthorn (*Crataegus*) flavonoids may have potential to limit brain damage from stroke.

One particular plant *Ginkgo biloba*, has come to the fore , for its strong anti-oxidant effects. *Ginkgo* seeds are sold in Asian shops and boiled to remove toxic components such as ginkgolic acid before use. In recent years *Ginkgo* extract has achieved a reputation as a treatment for dementia but may have wider role in limiting apoptosis. Unfortunately perhaps although Ginkgo biloba extract does enhance attention and memory in young healthy subjects at first, they develop tolerance to it so that the effect is lost after 6 weeks . Ginkgo extracts also have a broad anti-microbial effect against several potentially disease causing bacteria such as *Klebsiella pneumonia*, *Pseudomonas aeruginosa*, *Staphylococcus aureus*, *S. epidermidis* and *S. pyogenes* , and is helpful in the treatment of peripheral arterial disease .

There is a great deal of unrealized potential in plants. In one study extracts from leaves and fruits of three Brazilian *Hyptis fasciculata* (Lamiaceae) and the palms *Copernicia cerifera* and *Orbignya speciosa* have proved to have a better anti-oxidant activity than ginkgo. And in Ayurvedic medicine there is *Evolvulus alsinoides* as an alternative to *Ginkgo* for memory enhancement, antiepileptic and immunomodulatory properties .

Antiseptics, antimicrobials and antifungals

Predicting the efficacy of a particular chemical compound can be difficult. There are very many different kinds and many plants produce complex mixtures of them. Nevertheless their therapeutic effect can be predicted to some extent by the functional groups. For example the alcohols, compounds which have a hydroxyl (-OH) group is attached to a hydrocarbon skeleton, including the phenolic alcohols where the -OH is attached to a benzene ring, frequently they show an antiseptic, antimicrobial or antifungal effect.

The simple phenols, aromatic alcohols (a hydroxyl group attached to a benzene ring) are valued as anti-bacterial agents. TCP trichlorophenol is a manufactured one but they are also extracted from some plants. One called apocynin, extracted from the roots and rhizomes of the Himalayan herb *Picrorhiza kurrooa* is effective in alleviating the symptoms of emphysema and has been used for a long time in Ayurvedic medicine for a number of disorders including constipation. Arbutin gets its name from Arbutus, the strawberry tree, but is found in other members of the heather family (Ericaceae), including *Arctostaphylos uva-ursi*, the bearberry, and also the leaves the pear, *Pyrus communis*. The efficacy of bearberry extract for the treatment of urinary tract infections (cystitis, urethritis), prostates) arises because in the alkaline conditions of the urine arbutin is hydrolyzed to the antibacterial phenol called hydroquinone. Arbutin, for example marketed in the UK in a bearberry extract loaded cream called Dermabrite©, is especially popular in some parts of Asia as a skin lightener: it inhibits the formation of melanin by inhibiting tyrosinase activity. *Vaccinium oxycoccus*, *V. macrocarpon*, the cranberry and *V. corymbosum* the blueberry are in the same family Ericaceae and their juice is also recommended for urinary tract infections . However this efficacy is related to condensed tannins that limit the adhesion of bacteria to the cells lining the bladder and urethral tract. Arbutin also has anti-oxidant anti-aging activity.

Eugenol is a phenolic compound, a phenylpropanoid with a free hydroxyl group. It is widely distributed in sweet calamus (*Acorus calamus*), tarragon (*Artemisia dracunculus*), citronella

215

(*Cymbopogon nardus*), Ylang ylang (*Cananga odorata*) huon pine oil (*Dacrydium franklinii*) and carrot seed oil (*Daucus carota*) among others. In earlier times sweet calamus was harvested to be strewn on the floors of houses for its sweet smell that also kept insects such as fleas down. Eugenol is especially in several spices such as cloves (*Syzygium aromaticum*) and a major component of bay leaf (*Laurus nobilis*), nutmeg (*Myristica fragrans*) and allspice (*Pimenta dioica*). It has antimicrobial and anaesthethetic properties, hence the use of clove oil in dentistry. Other related compounds are Thymol and carvacrol are monoterpene phenols found in thyme (*Thymus*) and oregano (*Origanum vulgare*). The Australian myrtle *Backhousia myrtifolia* is a rich source of the phenylpropanoids essential oil (elemicin and methyl eugenol).

Other antimicrobials are often alcohol like menthol, terpinen-4-ol and geraniol. Terpinen-4-ol is particularly valuable and is found in such diverse plants as marjoram (*Marjorana hortensis*) and tea-tree *Melaleuca alternifolia* and *M. linearifolia*. Tea-tree oil is a particularly favored source of terpinen-4-ol. An Australian standard for it requires 30% of another component α-terpineol. Tea-tree oil is particularly favored because of its effectiveness for a wide range of skin problems and its relative lack of adverse reactions. Its wide range of activity – anti-microbial, anti-fungal, anti-parasitic may result from components apart from terpinen-4-ol and α-terpineol, such as the terpene hydrocarbons pinene, terpinene and cymene, sesquiterpenes, and other compounds like cineole. Over 100 different terpinenes have been identified in tea-tree oil.

For coughs and colds

Perhaps the commonest and most regular diseases humans experience are those of the respiratory tract, the coughs and colds. Needless to say a wide range of remedies have been proposed. *Astragalus*, garlic, cayenne pepper, ginger, lemon balm, mint, yarrow and lemons all have their advocates. *Astragalus membranaceus* Huang Chi is a traditional Chinese remedy rich in active compounds of which the saponins and complex polysaccharides may have the immune-stimulant effect. Clove powder is an essential ingredient of 'Composition Powder' made by

Samuel Thompson, an favorite old American remedy for the common cold, where it is mixed with bayberry bark, powdered ginger, white pine bark and cayenne pepper – certainly several components to make you feel warm at least!

One should really distinguish between those plants that relieve symptoms and those that fight the virus either directly or by stimulating the immune system. *Eucalyptus* and peppermint relieve nasal stuffiness. Sage and yarrow soothe sore a throat. Marshmallow. Coltsfoot, mullein relieve mucilage. Many different plants are expectorants, clearing the air passages and thinning the mucilage allowing an easier cough. Guaifenesin (guai or 1,2-propanediol, 3-(2-methoxyphenoxy) is an ingredient of many cough medicines. It is a synthetic compound derived from guaiac from the lignum vitae tree *Guiaicium*. Guaiac is a resin that is absorbed relatively rapidly and thins the mucus in the air-passages to make it easier to cough and clear them. It used to be used with mercury to treat syphilis. Anise and Mullein, in tea form are both natural expectorants.

Anti-virals, at least in test-tube studies, are barberry, elderberry, goldenseal, goldthread, horseradish, myrrh, Oregon grape, and indigo. The essential oils like menthol also may have an ant-viral effect. Several of the immune stimulants have already been mentioned. They include *Panax ginseng*, boneset (*Eupatorium perfoliatum*), Siberian ginseng *Eleuthrococcus senticosus*, garlic, indigo and *Schisandra*. *Hydrastis canadensis* (goldenseal), hyssop (*Hyssopus officinalis*), linden (*Tilia cordata*) are others.

Echinacea is one of the few herbal remedies that has become part of every-day commerce in the west. Tablets can be found on supermarket shelves. And yet the evidence for the efficacy of Echinacea against the common cold is still equivocal. Hundreds of trials have seemed to demonstrate that it is effective against the common cold but they have often been criticized for their lack of proper scientific design or poor statistical analysis . Nevertheless the weight of evidence does indicate the efficacy of *Echinacea*. This efficacy may be due to any of the three classes of immune-stimulatory compounds it contains; high molecular weight polysaccharides, chicoric acids and alkylamides. The high eight polysaccharides are particularly interesting. They are also present in cinnamon often a component of cold remedies in traditional

Chinese and Ayurvedic medicine. Cinnamon bark contains bioactive polysaccharides that are not digested but seem to stimulate the immune system via the Peyer's patches in the intestine. In Chinese remedies cinnamon might be combined along with Ma Huang (*Ephedra*) to induce sweating, or *Bupleurum falcatum* for its anti-inflammatory saikosaponins. In Ayurvedic medicine with Tulasi, Licorice, Ginger, Peppermint and clove among others! The kind of synergy between all these components is very hard to study.

Another traditional remedy is *Andrographis paniculata* also called Indian Echinacea, it is mainly distributed in India where it is part of the Ayurvedic tradition and called Kalmegha. In traditional Chinese medicine it is called Chuan Xin Lian, Its main active components, called angrophalides, are diterpene lactones. As well as having an immulo-stimulatory effect it forms part of a bitter medicine called *alui* that is a true panacea. It Nearly 30 different activities have been reported and every year more and more scientific papers have appeared describing its chemistry and therapeutic value.

Aromatherapy

Some of the alcohols and phenolic compounds mentioned are essential oils; compounds that are relatively low molecular weight that give plants their characteristic smells. They dissolve readily in alcohol but less readily in water, evaporate relatively easily, and importantly for their therapeutic value are absorbed easily by the body. Hence they have become the basis of aromatherapy.

To the outsider aromatherapy appears to be just another gimmick, another part of the New Age counter culture. A survey of the internet reveals enough exaggerated claims and gobbled-gook to put anyone off. Oils are divided into top notes, middle notes and base notes as if the therapist is manipulating the recipient like playing a musical instrument by assaulting his olfactory senses. Basil, bergamot, clary sage, coriander, eucalyptus, lemongrass, neroli, peppermint, sage and thyme are all short-lived top notes and used to stimulate and uplift. Middle notes are balm, chamomile, fennel, geranium, hyssop, juniper, lavender and rosemary and said to effect body functions and metabolism. Base notes are longest lasting and calming and comforting. They include cedarwood, clove, frankincense, ginger, jasmine, rose, sandalwood and rosemary.

And yet the oils used do have in some cases proven ability to alter mood as well as a therapeutic effect. For example Lavender oil contains the alcohol linalool and more than 100 other components. Linalool is a major component of mint *Mentha arvensis* essential oils too and is widely distributed in such plants as basil, thyme and cardamom. As well as having an antimicrobial effect linalool has a calming effect on the central nervous as a sedative and a local anesthetic.

Seeking the cure for cancer

No other area of the search for new drugs has raised such hopes, so often unrealized, than the search for the cure for cancer. This, of course, supposes that cancer is a single disease rather than a collection of miscellaneous diseases though characterized only by the uncontrolled proliferation of cancerous cells. In fact several plant based drugs are antineoplastic, that is they show a marked potential for limiting cell division, and thereby the growth and spread of cancerous tumors.

Vincristine is one of what are called indole alkaloids, alkaloids based on the indole molecule a heterocyclic molecule of a 5 carbon and 6 carbon double ring. *Catharanthus roseus* the Madagascar periwinkle is rich in alkaloids, over a 100 have been detected, including vincristine. In testing it for the treatment of diabetes it was discovered to retard the progress of leukemia., acts by preventing cell division in tumors where unregulated cell division is taking place (antineoplastic) and is used in the chemotherapy of childhood leukemia and Hodgkin's disease. Others do have a hypoglycemic effect. The common periwinkles such as *Vinca major* also contain indole alkaloids such as vincamine, majdine and majoridine that are antihemorrhagic and astringent but do not limit tmor development.

Taxol, is another drug that prevents cell divison and so helps to halt the growth of of tumors. It interferes with the development of the mitotic spindle, the network of tubulin microtubules that makes sure the chromosome divide properly. Taxol is mainly harvested from *Taxus brevifolia* the Pacific yew though this has become quite rare from over-harvesting. Taxol is a complex diterpene.

The compound indirubin is the active component of a Chinese

herbal remedy called Dang Gui Long Hui Wan, though this is the usual mixture of plants found in traditional Chinese medicine . Indirubin-3'-monoxime comes from *Indigofera tinctoria* (Indigo). It arrests cell-development before it enters the mitotic division stage (G" stage) and has proved useful in the treatment of chronic myelogenous leukemia (CML). Other *Indigofera* species have a variety of therapeutic effects protecting against liver damage, as an anti-inflammatory, and to treat diabetes.

One of the compounds in celery *Apium graveolens* that gives it its characteristic smell and taste is 3-n-butyl phthalide. This and another celery component called sedanolide have been shown to reduce the incidence of tumors in laboratory animals and in chemically induced liver cancers in tissue cultures . Highest concentrations come from the seed oils. Another component of the seed oil perillyl alcohol has also been associated with tumor regression in the pancreas, colon and lungs in animal studies. It prevents the blood vessel development, abnormal angiogenesis, that is essential for hard tumor growth. It is the perillyl oil component, among others in the crude seed oil extract, and also found in *Conyza newi*, an East African herb, that shows an insecticidal effect, fumigant and an anti-bite effect against mosquitoes .

The adaptogens - ginsengs and others

In different cultures there are special plants that are valued as adaptogens. They do not treat any specific illness but by taking them the body is brought into harmony and inner peace is achieved. Several plants that have already been mentioned that appear to be cure-alls but the adaptogens are different they have a generalized effect increasing vitality and helping the body cope with stress. The foremost adaptogens are the ginsengs. The name is a misnomer because there are several ginsengs, not all related to each other.

The efficacy of Ginseng root, *Panax ginseng* was advertised in part because of the root to the human body. It is perhaps the most famous of all Chinese medicines is, but paradoxically its efficacy against anything much has been doubted despite laboratory and clinical evidence for immune system modulation, anti-stress activity, and anti-hyperglycemic activity. There are three species, Chinese *Panax ginseng*, Korean and Manchurian *P. pseudoginseng*, and American *P. quinquefolius*. The root contains a wide range of constituents, 2-3% of which are 'ginsenosides', a kind of saponin. Saponins are a class of triterpenoids. The saponins form a collloid in water, and lather up like a detergent because they have fat soluble (lipophylic) portion at one end of the molecule and a water soluble (hydrophilic) portion at each opposite ends. There are approximately 28 different kinds of ginsenosides in the root of ginseng. They have a range of effects. Paradoxically it is its use as a universal panacea, as a general tonic for bringing the body into "harmony", as a tranquilizer and a sedative, a treatment for insomnia, agitation and anxiety, as well as for improved digestion, that detracts from the scientific valuation of ginseng.

The adaptogenic activity in Siberian ginseng (*Eleuthrococcus senticosus* or *Acanthopanax senticosus*) has long been recognized in traditional Korean medicine. It comes from the same family as the herb *Panax* but it is a shrub. Extracts of the roots and stem bark are used as a tonic to strengthen the qi. They are neuroprotective effect by inhibiting inflammation and the activity of the cells in the brain that first respond to damage the microglial cells. Its phenylpropanoids are similar to brain chemicals, catecholamines, such as epinephrine and L-dopa. Apparently Russian athletes used it as a tonic at the Moscow Olympics in 1984. Another plant with a similar chemical constitution is roseroot (*Rhodiola rosea* - Crassulaceae) which has a proven anti-fatigue effect .

What many of these plants share is their chemistry, the presence of phenylpropanoids, triterpenoids, saponins and related compounds that share a chemical structure similar to the steroids found in animals, which may lie behind their effects. For example the yam *Dioscorea villosa* produces the steroidal saponin diosgenin that is used as a precursor in the synthesis of several human steroids including the sex hormone progesterone used in birth control pills.

Triterpenoids, an important class of active compounds that are based on a four ring skeleton. Phytosterols are one class of triterpenoid. They inhibit the growth of tumors but have gained an increased significance because they compete with cholesterol in the diet. Obtained from oat-bran and soya among other plants several spreads and oils are now marketed as an effective way of reducing blood cholesterol level. Phytosterols called the guggulsterones are found in the resin myrrh, *Commiphora mukul*; they stimulate the thyroid and thereby reduces blood cholesterol level and triglycerides . Another kind of phytosterol are withanalides, steroidal lactones or triterpenoids with a lactone attached, that are important constituents of ashweganda or winter cherry (*Withania somniferum*) that is held in high regard in Ayurvedic medicine as a panacea, and is sometimes called the Indian ginseng. Ashweganda means "strength of ten horses". It is a mild sedative and an adaptogen, helping the body cope with stress, and an overall tonic.

In India there is another herbal adaptogen of great repute. This is Tulasi (*Ocimum sanctum* or *O. tenuiflorm*). It is widely cultivated and worshipped . In mythology it arose from the ash of Tulasi, the epitome of womanly virtues, but the wife of the demon Jalandhara. Vishnu tricked Tulasi into unfaithfulness in order to weaken Jalandhara so that he could be killed and then Tulasi committed sati as an act of faithfulness to her husband. The leaves and seed are eaten for a wide range of problems; heart and blood diseases, asthma, bronchitis, vomiting, lumbago, and fevers. It contains terpenoids with proven antibacterial, anti-fungal and anti-viral activity. Perhaps more central to the high value in which it is held are its ability to enhance the immune system and as an adaptogen that may come from its phenylpropanoid content. The phenylpropanoids have a structural similarity to the catecholoamines like dopamine and the circulating catecholamines epinephrine and norepinephrine. They prepare the body to meet emergencies such as cold, fatigue, and shock. Parkinson's disease is caused by a deficiency of dopamine in the brain. Epinephrine stimulates the heartbeat and is used to treat allergic reactions such as emphysema and bronchial asthma.

Glycyrrhetic acid is a triterpenoid saponin from licorice root (*Glycyrrhiza*) that acts to enhance the activity of endogenous steroids. The glycrrhetic acid is produced in the intestine by the hydrolysis of glycyrrhizin or glycyrrhizic acid, a glycoside that is 50 times sweeter than sugar, that is found in the roots and rhizomes of licorice . Licorice is well known as an aid to digestion, and is also used in cough remedies as an expectorant, though today its major use is in confectionary. Another flavoring and adaptogen is sarsaparilla that comes from the rhizomes of American species of the climber *Smilax* such as *S. regelii*, *S. aristochiifolia* and *S. febrifuga*). From the 16th century it gained a reputation as an effective treatment for syphilis, which puts into a different context the predilection for it of cowboys of the Wild West. It has a range of saponins (mainly sarsaponin, smil-saponin and sarsaparilloside) and phytosterols such as beta-siterol. A wide range of therapeutic effects have been claimed from sports performance enhancing to the treatment of skin diseases, kidney disease and rheumatism but most are as yet unsubstantiated.

Ziziphus jujuba provide another kind of triterpenoid saponin, called jujuboisides, that are another component of cough remedies. A North African species *Z. lotus* provides the lotus fruit that brought happy forgetfulness and indolence to the lotophagi, the lotus eaters, of North Africa. As Tennyyson put it,

"In the hollow Lotos-land to live and lie reclined
On the hills like Gods together, careless of mankind."

The lotus tree was one of the standard garden plants of the Ancient Egyptians. In Chinese the fruit is called *suanzaoren*, meaning "sour date seed". It is the most important component of a combination, Suanzaoren Tang that that is remedy for insomnia and fatigue, and includes as well as *Zizyphus* other sources of triterpenes, licorice and the immuno-stimulatory fungus *Hoelen* (*Poria cocos*), as the lily *Anemarrhena asphodeloides* and the umbel *Selinum*. There is also a well-known sedative combination of *Zizyphus* with wheat and licorice called Dazao Tan.

Several plants have been used to treat anxiety and depression(anxiolytic). *Hypericum* (St John's Wort) has shown efficacy for treating less serious forms of depression . Valerian acts, like the benzodiazepine tranquillizers such as diazepam valium, by enhancing the activity of the nerve transmitter GABA (gamma-aminobutyric acid) at GABA-a receptors, which dampens the brain's arousal system. About 40% of the synapses between nerve cells in the human brain have GABA receptors. When GABA binds to them they change shape slightly allowing negatively charged chloride ions to enter the neuron thereby reducing its excitability. Valerian contains multiple chemical constituents that act synergistically, such as volatile oils sesquiterpenes and monoterpenes, valepotriates, alkaloids, and lignans; it is the monoterpene component that mediates this effect. Valerian was used as a treatment for epileptic convulsions . Other herbs that have benzodiazepine activity are chamomile and *Passiflora incarnata*.

Who owns the drugs?

The drug companies are some of the biggest, most highly capitalized and most highly profitable companies in the world. They are multi-national conglomerates and towering examples of globalization. They invest hugely in the discovery, development and marketing of new drugs. Human health and well-being has been greatly aided by their products. And yet there is a serious concern about the way in which the use of drugs in therapy has shifted from the individual and the local practitioner utilizing local herbal remedies to a situation where faceless multinational conglomerates by deciding what to research, largely on the basis of profit, also decide the health priorities of the population. The discovery of many of the drugs that finance the drug companies is related to the use of particular plants in traditional medicine and yet they have usurped traditional medicine.

A particularly shady aspect of this is bio-piracy where living resources or traditional knowledge and practices predominantly in developing countries are patented so that their development becomes the exclusive privilege of companies in rich developed companies whose priorities are profit led. Many countries, like India, have strict laws to defend their "biological heritage" from the piracy of western nations, though it is possiblethat these also act to inhibit scientific research.

The patenting of the products of the Neem tree has brought this issue sharply into focus. The Neem tree (*Azadirachta* indica). It has been used for hundreds of years in India and Burma and is a veritable cure-all. Chewed neem twigs keep teeth healthy and neem extract, containing salanin, is now included in some toothpastes. Neem oil contains a range of compounds most derived from the tetracyclic triterpenoid Tirucallol including Azadirachtin, Azadiradione, Nimbin and Salanin . Tirucallol is also the active component in the traditional Chinese medicine Kansui root made from the powdered root of *Euphorbia kansui*. Neem oil repels insects of many sorts including mosquitos, chiggers and ticks, lice and scabies and is also fungicidal, bactericidal and anti-viral. The anti-insect effect relies on several of these acting together but with a different target. Neem oil has also been used as a contraceptive douche!

Azad-Darakth is the Persian for "free-tree" so that it is highly ironic that a long court case has been fought over who owns the tree. On 8th March 2005 after a 10 year legal debate the European Patent Office made a historic decision to uphold its previous decision in 2000 to cancel their patent on the fungicidal properties of seeds extracted from the neem tree first granted in 1994 to the United States Ministry of Agriculture (USDA) and the chemicals multinational WR Grace. The campaign against the patent had been run by the Indian environmentalist Vandana Shiva, Magda Aelvoet, MEP and formally President of the Greens in the European Parliament, and the International Federation of Organic Agriculture Movements (IFOAM).

This seems a very straightforward ethical decision and yet the devil is in the detail. The patent of WR Grace, there are many others associated with the Neem tree, is actually of a new technological process that extracts and stabilizes the active component aza A from neem oil and extends its shelf-life to two years or more . It is neither a patent of the actual plant nor of its genes. The original campaign against the WR Grace patent was made on the basis that biological resources are common heritage and should not be patented, the patent will restrict the availability of living material to local people whose ancestors have spent centuries developing the material, and the patent may block economic growth in developing countries none of which seem to be at all justifiable in this case. Other patents exist including three Indian ones but *Azadirachta* is not just an Indian plant but is widespread in South and Southeast Asia and may even have originally been native to Burma. So whose plant is it?

At least when the major drug and food companies market a product there is a degree of quality control. They have a reputation to maintain. An example of what can happen is the marketing of *Hoodia gordonii*, a miracle appetite suppressant. It contains a steroidal glycoside called termed P57AS3 (P57). Studies have shown that it appears not to act by binding known CNS (central nervous system) receptors but increases the content of ATP in hypothalamic neurons by 50-150% . *Hoodia* is a rare cactus that grows in southwest Africa, in South Africa and Namibia and is used traditionally as a food by the San. It is an endangered species in the wild, but there are a very small number of farms in south Africa licensed to grow it and export it under C.I.T.E.S. (Convention on International Trade in Endangered Species of Wild Fauna and Flora) certification. One consequence of this is that there is a flourishing internet scam trading in tablets and capsules purporting to contain Hoodia but in fact containing either a Chinese herbal mixture or some other "herbal" materials and other adulterants.

7 Plants For The Soul

"Welcome the Creations Guest,
Lord of Earth, and Heavens Heir.
Lay aside that Warlike Crest,
And of Nature's banquet share:
Where the Souls of fruits and flow'rs
Stand prepar'd to heighten yours." - Andrew Marvell

There is so close connection between the use of plants for health and their use for recreation and pleasure that it is a very artificial to separate them as I have in this book into two chapters. These plants and their products range from the mundane, the cup of tea, the sugary sweet, the flavor-some spice, to the extraordinary hallucinatory or psychedelic drugs. The use of plants that supply human needs, body and soul, has had a profound influence on human history. The pleasure plants contain compounds with a physiological effect on humans, some are panaceas, most are poisons if used in high enough concentration, and most are psychoactive. The more powerful the medicine or the recreational drug, the more strongly the association with myth and religion, the more commonly they have been used in religious ritual, as a means of connecting with a spiritual world.

The story of drug use is very complex. Attitudes to drug taking have been markedly different at different times. In the modern world our attitudes to drugs and the plants that make them is a particularly artificial one, one that arises from a peculiarly Judeo-Christian-Islamic tradition that separates the body and the soul. The Jewish tradition is opposed to the use of most drugs except for those which are medically prescribed. Perhaps the earliest Christian document the Didache commands people not to use the confections of sorcerers. This is a relatively recent construct and has grossly distorted our attitudes to "drugs". It has given rise to some absurdities such as the distinctions, between plant products with psychoactive components that aren't regarded as drug plants at all, legal stimulants and illegal drugs, and between "soft" drugs and "hard". Legal systems have sought to define the boundaries of acceptability and the levels of punishment if those boundaries are exceeded by categorizing drugs. In the UK by the Misuse of Drugs Act 1971 drugs are categorized as Class A, B or C depending upon how harmful they are considered to be. Opium, morphine, heroin, and cocaine are class A drugs, Codeine is a class B drug. Cannabis is currently a class C drug. Alcohol, the drug that probably causes the most widespread social damage is not included.

The US Department of Health and Human Services categorizes drugs as Narcotics, Depressants, Stimulants, and Hallucinogens, among others categories, but has a difficulty categorizing Cocaine or Cannabis. Narcotics relieve pain, may provide an initial rush and feeling of euphoria but then deep relaxation and loss of anxiety. Stimulants provide the euphoria without the relaxation. Depressants provide the relaxation and loss of anxiety without the euphoria. Alcohol, in its own special category, maybe because it is not illegal at present, is a depressant but at a certain stage of the evening, in the right company acts like a stimulant. Down the ages users of all sorts of drugs have noted something similar; the effect of a drug depends upon the mood of the individual and the circumstances. Cocaine acts as a powerful stimulant when it is taken in powder form as a kind of snuff, but in its traditional use as leaves held between the cheeks and gums and sucked it acts as a narcotic. Similarly cannabis as marijuana has a variety of effects, euphoria, hilarity, loquaciousness, nausea, paranoia, even no effect, depending upon mood, but when more concentrated as hashish produces a strong psychoactive, even a hallucinogenic effect. Euphoric effects akin to those reported for stimulant drugs have even been reported for nicotine, but it is not on the Department of Health list of drugs. Neither is caffeine, another stimulant.

In 1924 a German toxicologist called Lewin published a slightly different system for classifying narcotic and stimulating drugs:
euphorica – sedatives (morphine, heroin, cocaine);
hypnotica – inducing sleep (kavaine);
excitantia – stimulants (caffeine, nicotine, betel nuts);
inebriantia - excitation followed by depression (alcohol)
and phantastica – hallucinogens (mescaline, atropine).

While most recreational drugs are psychotropic or psychodynamic, affecting the mind, emotions, and behavior some also profoundly disturb normal perception, cause hallucinations, and produce effects analogous to psychoses like schizophrenia.

A very broad range of plants have been used around the world for the stimulant or excitatory effect or as intoxicants or as hallucinogens. In each region a different range of plants has been utilized. For example the Masai in Kenya utilize 20 different plants. One is olkiloriti or *Vachelia nilotica*, an acacia tree that is widespread in Africa, whose bark and roots are said to have given the famous Zulu warriors courage before their battles. However many of these native sources have been supplanted over time by a much smaller range of plants that now dominate the world of recreational drug use: coffee, tobacco, cannabis, opium, and cocaine. For example in much of Africa "Dagga", Cannabis sativa is now so much a part of the native culture that it seems to have originated there, but it was only introduced relatively recently, probably by Arab traders. Widely valued medicinally Cannabis sativa in places it also became part of a social activity, with people clustering around a small hole and taking it in turns to cup their hands over it to inhale the smoke from a mixture of burning dagga and dung.

There are many strands to the human story of recreational drugs. Not least is the nature of the chemicals and their different effects on the human brain, but there is another story, one of the changing social context and changing means by which plant drugs have been taken. Over time there has been a shift from ingesting the plant, either chewing it or drinking a tea made from it, through the use of snuff, to the inhalation of smoke, to its injection. Each stage has multiplied the immediacy and magnified the effect of the plant.

The Pleasure Principal

Various different recreational drugs act upon the reward circuit in the brain where they increase in several different ways the amount of dopamine available. In chapter five the complex action of alcohol on the reward circuit was described. Different recreational drugs mediate their effect at different places.

Caffeine is an alkaloidal amine, a methylxanthine. It binds to the adenosine receptors. Adenosine acts as a natural brake on neural activity when it binds to its specific receptors on neuron synapses, slowing down their rate of response to neurotransmitters, making you feel sleepy. By preventing adenosine binding caffeine enhances neural activity. This extra activity leads to a greater release of adrenaline by the pituitary increasing energy and concentration. Caffeine also acts to increase activity of dopamine in the brain's pleasure/reward circuit because adenosine acts as a negative competitor, an antagonist at dopamine receptors D2. When caffeine reduces the binding of adenosine at the dopamine receptors, dopamine activity and turnover is enhanced. In turn dopamine acts as an antagonist at adenosine receptors A2, reducing the braking activity of adenosine in the brain, increasing motor activity.

Nicotine acts both to increase the activity of dopaminergic neurons and the amount of dopamine. It is an alkaloid with a nitrogen in the six-membered ring, properly called 1-methyl-2 (3 pyridyl) pyrrolidine. It mimics the neurotransmitter acetylcholine on sodium uptake channels on dopaminergic neurons but without the normal temporary desensitization after activation, thereby leading to a cascade of dopamine in the *nucleus accumbens*. Nicotine also inhibits an intracellular enzyme called MAO B (monoamine oxidase B) that breaks down dopamine after its reuptake. As a result there is a general rise of the amount of dopamine.

The active ingredient in cannabis is delta-THC, Δ9-tetrahydrocannibol. Unlike most psychoactive drugs it is not an alkaloid but a lipophyllic resin. The other exception are the kava lactones. While many alkaloids are toxic in high concentration, that is they will kill, cannabinoids have a low toxicity. The long term effects of different drugs vary too. Some are cause physical and /or psychological dependence. Others do not. Long term use can be life- and lifestyle damaging or not; it depends upon the individual and their circumstances. Some have unfortunate long term effects for some users; for example, it is increasingly clear that use of high THC content cannabis products is linked to the onset of bipolar disease (schizophrenia). The therapeutic use of cannabis in managing chronic pain and multiple sclerosis has become established in recent years. However it has also become clear that cannabis use is associated In some users with psychiatric disorders such as dependence, anxiety, depression, cognitive impairment, and psychosis.

THC concentrates chiefly in the ventral tegmental area and the *nucleus accumbens*, but also in the hippocampus, the caudate nucleus, and the cerebellum. It produces many different effects . In the same way that opiate drugs mimic the endorphins, delta-THC mimics an endogenous molecule called anandamide. Anandamide receptors, called CB1 receptors are widespread throughout the brain. If delta-THC binds with them the receptors interact with several intracellular enzymes including cAMP with the result that another enzyme called protein kinase A is present in lower concentration in turn reducing the activity of potassium and calcium channels. The net result is reduced neural activity and less GABA is secreted. Alternatively dopaminergic neurons lack the CB1 receptors but have GABA receptors that inhibit them in the presence of GABA. Because there is less GABA in the synaptic region the dopaminergic neurons show greater activity and there is a cascade of dopamine.

Opiate drugs are isoquinoline alkaloids. Opium, the dried latex of the opium poppy (*Papaver somniferum*) contains over 40 different alkaloids. The most important is morphine. Heroin is a synthetic derivative of morphine (see below). Opiates act in a similar way to nicotine, that is by mimicking neurotransmitters /neuromodulators, but this time the neurotransmitters are the endorphins (i.e. endogenous morphine). Morphine itself is actually produced endogenously produced by neuroblastoma cells but other endorphins include beta endorphin. The ventral tegmental area is targeted by opiate drugs like morphine. as well as mimicking other endogenous endorphins. G-protein coupled receptors mediate opium effects, interacting with many different proteins and modulating many signaling molecules . Morphine interacts with the brain in a complex way. It not only acts on the ventral tegmental and the nucleus accumbens but also but also on other parts of the brain that are influenced by endorphins such as the amygdala, the *locus coeruleus*, the arcuate nucleus, and the periaqueductal grey matter, indirectly altering dopamine levels. Morphine is one of the strongest analgesics probably because of its effect on the thalamus.

Cocaine is tropane alkaloid. It acts very directly on dopamine levels by targeting reuptake pumps. For example it targets the D1 receptors on dopaminergic neurons. Here it prevents reuptake dopamine after its release, thereby it increases its concentration in the synapses. A feeling of euphoria arises from the cascade of dopamine in the ventral tegmental area and nucleus accumbens. Cocaine also prevents the reuptake of serotonin and norepinephrine. Similarly cocaine induces a feeling of confidence by increasing the levels of serotonin that are present. Cocaine has a subsidiary effect on the caudate nucleus that is associated with stereotypical behavior and repetitive behavior like nail-biting. Unfortunately, in chronic cocaine user neurons adjust, by reducing the number of dopamine receptors for example. Without the constant supply of cocaine to maintain the higher levels of dopamine that are now needed for normal behavior there are cravings and withdrawal symptoms such as depression.

Beverages

Plants have provided recreation. Teas of various sorts made from infusing leaves or roots with boiling water, have been used for millennia. As well as extracting the chemicals from the leaves boiling water had the benefit of sterilization when water sources were likely to be polluted. There is no sharp distinction between using an infusion of a medicinal plant for its therapeutic purposes and just for pleasure. Most of the plants used for beverages have a proven therapeutic value.

Mint (*Mentha*) tea is popular in Islamic cultures. The mints contains essential oils menthol, menthone, menthyl acetate flavored with many other minor components and is good for indigestion and colic. There are several species such as peppermint *M. piperita*, spearmint *M. spicata*. They are rich in essential oils but much of their pharmacological activity is related to menthol. It cools by desensitizing warmth receptors. It has been especially valued as an aid to digestion against abdominal pain, flatulence, and indigestion.

Rooibos (redbush) tea made from *Aspalathus linearis* from South Africa. It is favored by some because it lacks caffeine. The color of redbush tea is like tea but unfortunately it also lacks the flavor-some tannins and to me tastes of boiled grass clippings. Nevertheless it has a high anti-oxidant activity and is said to ease stomach cramps and diarrhea and relieve eczema and allergies. Many other herbal teas are now produced with or without real tea to provide flavor:

chamomile to relieve stress, valerian to help you sleep, lemon balm to lift the spirits, and mint tea to ease stomach problems or headaches, are just some of them.

Valerian tea (*Valeriana officinalis*) is noted for its sedative effects probably induced by the iridoid alkaloids called valepotriates. They seem to act as a sedative by inhibiting the uptake and increasing the release of the inhibitory neurotransmitter GABA (gamma-aminobutyric acid). Valepotriates decompose readily, do not dissolve easily and are not readily absorbed so that valerian tea drinkers are generally exposed to very small quantities , which is just as well since at higher concentrations they produce a number of adverse reactions such as hepatotoxicity and hypersensitivity.

Pacific Islanders prepare Kava from the rhizomes of the shrub *Piper methysticum*. Its mixture of different lactones produce a feeling of brotherhood and tranquility when administered in the formality of the kava ceremony. For some Pacific islanders Kava drinking sessions have become a link to a mythologized and idealized past . Like the ritualistic drinking of wine a few drops are allowed to fall to the floor as a libation to the ancient gods. Kava drink used to be prepared by the women chewing roots, mirroring the chewing of grains used to start alcoholic fermentation in many parts of the world, but here it worked only physically to release the resins in the roots. Kava contains 5-10% resin rich in lactones called kava pyrones: the three most important components are kavain, dihydrokavain and dihydromethysticin derived from the alpha-pyrone 5,6-dehydromethysticin. The first anaesthetizes the mouth like cocaine but the second is more readily absorbed and metabolized. Kava is hypnotic, sedative and a muscle relaxant but is said also not to reduce mental awareness but it can become habit forming.

The Caffeine hit

The most widely used teas are stimulants. Most, including "tea" itself, contain caffeine. The caffeine content of the raw product does not really indicate the caffeine "hit" a beverage provides because of the different ways they are prepared and the strength of the cup. Typically a cup of instant coffee can have more than twice as much caffeine as a cup of tea brewed for 1 minute. Ground coffee has much more but so has stewed tea. Cocoa has a quarter the caffeine of tea, but chocolate more. Theobromine is a weaker stimulant than caffeine. It is also found in tea and mate. Colas of various sorts rival instant coffee for their caffeine content.

Like many other beverages humans rely on to assuage their souls, coffee provides a rich mixture of chemicals with complex physiological effect, some good some bad. As well as caffeine, coffee beans contain the phenylpropanoid, caffeic acid. Experimentally this inhibits the enzymes DOPA-decarboxylase, important in neuro-transmission, and 5-lipoxygenase, involved in the inflammatory response to shock , but more beneficially it also inhibits the growth of tumors in liver cancer and aids digestion. Green coffee beans contain up to 10% of chlorogenic acids, i.e., various isomers of hydroxy-cinnamoyl esters of quinic acid (a sugar-like molecule and a common plant constituent).

In the roasting process, some of the chlorogenic acids form esters by losing a molecule of water, forming non-acidic quinolactones (quinides). Brewing results in very many different compounds as the quinolactones isomerize. Although they may only be a small component by dry weight (0.3%) they readily enter the brain and bind with receptors altering mood .

Coffee is another Old World species probably originating as a crop in Ethiopia. Its oldest recorded use is the chewing of leaves or beans to relieve fatigue and hunger. Coffee drinking originated in Arabia surprisingly late and not until the 15th or 16th century did it become widespread, especially spreading from Turkey to the rest of the world. In 1551 Suleiman the Magnificant granted permits for coffee houses in Istanbul.

In the second half of the 17th century coffee and cocoa were widely available in coffee houses in the cities of Europe. London's first coffee house opened in 1652 by Pasqua Rosee. The Angel coffee house in Oxford had opened two years before. The coffee houses provided a new social venue in which business, new ideas and gossip could be freely exchanged. Soon there were hundreds of them. King Charles' mistress Nell Gwyn even appeared in a comedy entitled the eponymous Coffee-house. In London Garaway's in Change Alley was a favorite with Robert Hooke and Christopher Wren. It was here that tea drinking was introduced to

London.

The diary of Samuel Pepys recounts many visits to Miles' coffee house in Westminster. Other customers included John Milton and it was here that an amateur parliament "the Coffee Club of the Rota" met. The coffee house became the nurseries of British party politics. Later the Whigs frequented the St James' Coffee House and Old Slaughter's and the Tories the Cocoa Tree Coffee-houses became business offices. In 1688 Edward Lloyd opened a coffee house that became the place where insurance brokers met, renting booths to meet their clients. It was a business that started with an advert in the London Gazette offering "a reward of a guinea for information regarding stolen watches, claimable from Mr. Edward Lloyd's Coffee House in Tower Street."a reward for stolen watches.

Coffee cultivation was introduced to the east from plantations in Yemen, especially by the Dutch who established plantations on Java by the late 17th century. There were two main introductions to the New World, one via the Amsterdam botanical garden of a single plant sent from Java. This was also the source of a few seedlings sent to the Paris Jardin des Plantes that were then sent to Martinique. Another introduction was from Yemen via Réunion and then to the Caribbean. As a result the genetic base of New World coffee is very narrow and it is very vulnerable to leaf rust (*Hemileia vastratix*).

Over time more and more sophisticated methods have been developed to brew coffee, to get nice strong but not too bitter coffee, and to separate the coffee grounds from the liquid coffee. In time a coffee pot was perfected with a long spout arising from a round swollen center to drew liquid from an area away from the floating grounds on top and the sunken grounds below. Textile coffee bags made from cotton or hemp were used to hold the ground coffee. The Biggin Pot had a built in cloth filter. A water spreader pot had an upper chamber in which the boiling water was added so that it dripped down evenly on the coffee below. Percolators The vacuum pot was developed by Madame Vaissieux of Lyons. There were two chambers. Water was heated in the lower chamber and forced up into the upper chamber where the ground coffee was placed. After the pot was taken off the heat, the steam in the lower chamber condensed drawing back the liquid coffee into the lower chamber through a filter. The percolator uses water vapor bubbles to circulate hot water around over the coffee grounds. Plunge coffee pots allowed quick brewing and the quick separation of the grounds from the liquid. With the advent of electricity, thermostatically controlled drip brewers were invented. Expresso machines force heated water through the grinds picking up not just caffeine but many other components, such as emulsified oils in the process, to create the smooth expresso taste.

In the 18th century tea replaced coffee as the most popular hot beverage in Britain and its colonies. In legend the Chinese emperor Shen Nung discovered green tea in 2737 B.C.E. from *Camellia sinensis* when some leaves fell into his drinking bowel. The Portuguese probably introduced tea to Europe from the Far East. By the early 19th century a huge trade through the port of Canton, probably half to Britain, was in place. It was the tax on tea re-exported to America that was used as one excuse for the rebellion that led to war of independence of the United States *"no taxation without representation"*. Chinese civilization had been brought to a low by other westerners by the sale of opium, which was inextricably bound up with the tea trade. Introduction of tea plants from China to India had failed, but then Wallich, once director of the botanic gardens in Calcutta, identified a native Indian variety of tea plant in the hills of Assam which became the foundation of the Indian tea industry. Today tea plantations are found throughout the highlands of Asia and Africa. For a long time Indian and Sri Lankan tea was regarded as inferior in quality to Chinese tea. Green Chinese tea has the tannin epigallocatechin as its main flavor component tannin. The tannins in black fermented tea are oxidized and very complex. Different varieties, conditions of growth, and treatment after harvest, result in the subtle differences in taste and color: Assam, deep bronze and malty; Ceylon, pale gold and delicate; Darjeeling, light and fragrant; Kenyan coppery and strong; Lapsang Souchong, smoky. It is curious that some of the more specialist teas like Earl Grey, with its citrus and bergamot, arose firstly as teas adulterated with other plant material to make up bulk and weight before they were traded. Like many plants used for pleasure the drinking of tea became the center of a ritual in both Britain and Japan.

The soft fizzy drinks now rival coffee and tea in popularity. Many of these are rich in caffeine but not from coffee but rather from Kola tree (*Cola nitida* and *C. acuminata*). Coca Cola and Pepsi

Cola are the world leaders but there are many others. Kola originated in West Africa and is a relatively recent domesticate, within the last 1000 years or so. The nuts that provide the caffeine rich flavoring of many soft drinks are actually the embryos after the fleshy seed coat is removed. A guarana (*Paullinia cupana*) based drink was popularized in 19th century France but in recent years guarana has been widely added to alcopops to make stimulant or reviving drinks because of its high caffeine (3-6%) and theobromine content. It also contains tannins and saponins. It is used traditionally with cassava to make an alcoholic beverage in the Matto Grosso. It is ground into a paste with cassava and then dried as sticks that can be crumbled to make a beverage. It also had a traditional use as a fish poison.

Other caffeine rich drinks have a more regional popularity. Maté or Yerba Mate (Brazilian or Paraguay tea) *Ilex paraguarensis* provides a caffeine rich tea that is popular in South America. It makes a fresh invigorating drink very like green tea but with a slight minty after taste. Wild trees are also harvested, along with a few related species to supplement the plantation supply.

Chews and sucks

In different parts of the world different plants have been chosen as chews. Chewing gum, chicle, from *Manilkara zapota* was chewed in Central America. It was collected from the wild by chicleros. It contains no particularly active compounds but the process of chewing leads to the release of natural endorphins. The use of chicle based chewing gum was introduced to the USA and hence the world by Santa Anna, the Mexican hero of the Alamo. When he was down on his luck he tried to interest Thomas Adams in chicle as a raw material to make rubber. This was a failure but Adams developed flavored chewing gum as an alternative to tobacco and to paraffin based gums that weren't very chewy.

Khat is a shrub and tree *Catha edulis* that grows in East Africa and the Arabian Peninsula. The leaves and young shoots are chewed socially throughout that area, especially in the Yemen . It has a sweet taste and dries the mouth, dilates the pupils, increases blood pressure and heart rate. It used before prayer because it induces a feeling of euphoria and increases concentration. It reduces feelings of hunger and so acts to reduce obesity. It contains alkaloids that are natural analogues of amphetamines called cathinone and cathine, which is also found in Ephedra. Different varieties have a higher concentration of the more active cathionine; red khat gives a bigger hit than white. The cathionine decomposes rapidly after harvesting and so fresh leaves, so leaves wrapped in polythene or wet banana leaves and refrigerated, are preferred. Unlike many drugs there is no evidence of tolerance building up but some people do become dependent on it. In the US cathionine is scheduled as a Class I narcotic and so legally restricted but it is unlikely that leaves imported into the US have much of this present by the time they are sold. It is not legally restricted in the UK.

Betel nut is the seed of a palm Areca catechu. After alcohol, caffeine and , betel is the most widely consumed recreational drug. It is consumed in South and Southeast Asia and the Pacific Islands by at least 200 million people. A slice of the nut is "sweetened" with lime and wrapped in a piece of betel leaf (*Piper betle*) and chewed. Betel nut contains several active compounds including an alkaloid similar to nicotine called arecoline, a potent muscarinic cholinomimetics, that stimulates the nervous system and respiration, causes the face to flush and has a mild psychoactive effect. As well as relieving the astringent taste of the betel the lime, aids the release of the alkaloid, but along with betel-phenol from the nut and chavicol from the leaf also induces salivation. The red mouthed users, and red spit splattered public spaces are familiar sights from South Asia to the Pacific Islands. One of the least desirable results of the habit is a high rate of cancer of the mouth among users, it is the most common malignant cancer in Papua New-Guinea, a consequence of the lime and the presence of an essential oil called safrole in the Betel nut. Alternatively betel users use more harmful recreational drugs less and are less prone to psychoses like schizophrenia , this latter Betel nut alkaloids.

Another SE Asian chew is Kratom, *Mitragyna speciosa*, from the same family as coffee the Rubiaceae. The leaves are chewed after removing the central vein, or it is brewed into a tea. In low doses it acts as a stimulant but in higher doses it is sedative, analgesics, a treatment for diarrhea and it is also used recreationally. The opioid effect is due to an indole-alkaloid 7-hydroxymitrag.

In the Americas tobacco used to be the favorite suck of native

South Americans. Coca leaves from *Erythroxylon coca* are made into a wad that is kept between the gums and the cheek and moistened with saliva The leaves are often masticated with some lime, either from the ash of another plant some burnt sea-shells. Cocaine inhibits the neuronal reuptake of the neurotransmitter dopamine. In nature it seem that it inhibits the uptake insect neurotransmitter octopamine and so acts as an insect anti-feedant. Before the conquest coca leaves were used in religious rituals. Coca eases the headache and nausea associated with high altitudes. They were smoked to induce a trance-like state by sitting in its smoke. In the Inca Empire they were controlled by the aristocracy. A tax consisted largely of prepared coca breads that their lords and masters ate in large quantity. The Spanish conquest ended that. The European colonialists soon saw the commercial potential of it: plantations of coca spread and in by 1569 Monardes was reporting the increasing use of coca among lower class Indians. Although the Church tried at first to prevent its use it was widely and highly profitably traded; 8% of the European population in Peru were occupied with its trade . It was of particular use in assuaging hunger and suffering among the slave and serf population. Sucking the leaves only increase the sense of well-being and seems to have few of the disadvantages of the purified cocaine alkaloid. Soon the tithe on the coca trade came to be the single largest item of revenue to the Bishops of Cuzco and Lima.

Smoking and snuff

By smoking or the taking of snuff active compounds gain an immediate entry to the blood stream through the thin mucous membranes of the respiratory tract. Smoking has an age-old history. Pliny records the smoking of several materials including dried dung from an ox fed on grass. It was the Scythians who are first recorded as using cannabis for recreational purpose. They took steam baths placing cannabis seeds on heated stones.

The taking of snuff was an established habit in Europe. Shakespeare records aromatic powders of orris root (sweet flag-iris *Iris florentina*), chamomile (*Chamaelium nobile*) and white pellitory (*Anacyclus pyrethrum*) being passed around banquets. Other plants used were alehoof (*Glechoma hederaceae*) and sneezewort (*Achillea ptarmica*). *Achillea millefolium* has been used as a substitute for tobacco in Sweden). Tobacco snuff was combined with orris, cinnamon, cloves, fennel sage, bergamot and lavender.

Mayan pottery figures record smoking, probably of tobacco, though *Nicotiana tabacum* is only one of about 60 species in the tobacco genus and other species were utilized, chewed, smoked or as snuff for the alkaloid nicotine they contain. Like other Caribbean tribes the Taino of Cuba used tobacco in rituals, recreation and medicine. They smoked tobacco in pipes and cigars but also made cigarettes wrapped in palm leaves, corn husk or bark. They sniffed a mixture tobacco and coca leaf dust to relieve fatigue and chewed tobacco as a stimulant. The first European to try tobacco was Rodrigo de Jerez, whom Columbus sent to explore inland Cuba. He liked it so much that he took the habit home to Spain but was imprisoned by the Inquisition for being in league with the devil for puffing away. Its use was widespread in Central and North America. In 1535 Jacques Cartier found it being used by the natives of the Montreal region. Tobacco had been used for millennia, as snuff, or it was smoked, by the Aztecs as cigarettes made in reeds with sweet smelling resins, or as bundles of leaves rolled together and called by the Mayans sik'ar, or in pipes. Alternatively the name is said either to come from the rustling sound of cigars, like cicadas (cigarra), or from the little garden (cigarral) where tobacco was first cultivated in Portugal.

By 1531 the Spanish were cultivating tobacco on Santo Domingo and by the middle of the 16th century the Portuguese were cultivating it for export in Brazil. Its rapid spread around the world must be due to its remarkable ability to turn its subjects into addicts . It triggers the release of dopamine, a molecule associated with experiencing pleasure, it stimulates and relaxes. It effects this by mimicking the brain transmitter acetylcholine. It has wider effects too on the adrenaline pathways and on serotonin levels. The addiction it causes in insidious because in the longer term, in chronic smokers for example, the brain compensates by adjusting the number of brain transporters and spreading new acetylcholine receptors in different parts of the brain, so depressing the brain reward function, or in other words the feeling of pleasure, so that more nicotine is required to get the same hit, and only further smoking can alleviate withdrawal symptoms .

In Europe it was advertised as a panacea. Seville was the great entrepôt of goods from the New World and so it is not surprising that it was a Sevillean physician Nicolas Monardes who wrote the most influential books on New World plants entitled *Historia medicinal de Indias occidentales*. The second volume published in 1571 is largely a long eulogy about the many qualities of tobacco . Monardes described the way Indian priests cast dried tobacco leaves on fires and inhaled the smoke, through the mouth, or sniffed the smoke through a cane; apparently this induced ecstatic drunken behavior and the seeing of visions. Since tobacco is not a hallucinogen one wonders whether it was coca leaf dust included with the tobacco that was stimulating the euphoria. More soberly Monardes recommended the use of tobacco for headaches, chest complaints and worms!

Monardes work was quickly translated first into Latin by Charles D'Ecluse and then into English. The English translation by John Frampton was published in 1580 and then given the wonderful title

"Joyfull newes out of the new found worlde, wherein are declared the rare and singular vertues of divers and sundrie herbs, trees, oyles, plants, & stones."

In that title it is easy to sense the excitement of the botanical riches of the New World. Perhaps the most rapidly adopted New World plant was tobacco. The first book entirely on tobacco was Giles Everard's *'De herbe panacea'*, advertising its many merits published in Antwerp in 1587.

Snuff was being sold in the markets of Lisbon by 1558. Jean Nicot the French ambassador to Portugal, is credited with introducing the fashion of tobacco use to northern Europe in the 1560s by recommending tobacco snuff to the queen Catherine De Medici for her headaches. Portuguese and Spanish ships were carrying snuff and tobacco, and its cultivation, around the world. Seville became a centre for the production of cigars as we know them: the first cigars were made by the Spaniard Demetrio Pela. Here too as a by-product, cigarettes, waste tobacco wrapped in paper, were devised. It was a fashion among sailors and they introduced the habit of smoking pipes to England, but everywhere it very rapidly became part of the native culture. The making of cigarettes was carried to Turkey by sailors. In the colonies in Virginia smoking it in a long clay stemmed pipe became the fashion. William Camden reports the introduction of this fashion to England in 1586 by a party of Virginia colonists returning to England:

"These men who were thus brought back were the first that I know of that brought into England that Indian plant which they call Tabacca and Nicotia, or Tobacco".

But like any fashion it takes a celebrity to establish it and this celebrity may have been Sir Walter Raleigh. By 1596 a German visitor to London noted the English passion for smoking and in 1597 John Gerard describes the smoking of pipes of tobacco. Like nutmeg, tobacco became especially praised as a prophylactic against plague. The habit of smoking and snuff taking was widespread in men, women and children. At Eton College the boys were beaten if they forgot to smoke their pipe of tobacco in the morning. It had become a world-wide craze; a set of smoking implements dating from the Ming dynasty (i.e. pre 1573) is displayed in the China Tobacco Museum in Shanghai.

It is remarkable how the backlash against developed; what was to be feared except that it was the introduction of some despicable alien habit. In 1575 smoking was banned in all places of worship in the Spanish colonies. By 1600 Pope Urban VIII was threatening excommunication for smoking and using snuff in places of worship and by 1624 banned the use of snuff anywhere; the pleasure that snuff inspires and the sneezing it provokes was thought to be sacrilegious as being too like sexual ecstasy! In other places more extreme penalties were imposed. By 1612 the first attempts were being made in China to ban it; by 1634 planting, trading and using tobacco was punishable by decapitation; the death penalty had previously been imposed in 1617 in Mongolia. However the habitual use of tobacco snuff even became more popular in the 1650s, a habit that survived to the 20th century. In 1628 in Persia Shah Shefi executed two tobacco traders by pouring hot lead down their throats. In the Ottoman Empire tobacco was particular associated with the introduction of degenerate foreign habits so that Sultan in 1633 the tyrant Murad IV banned it. on pain of execution. Ironically alcohol was also banned though the Murad IV was a mad alcoholic who died of cirrhosis of the liver at the age of 27. In Russia in 1634 Tsar Alexis imposed whipping, slitting of the nose and exile to Siberia as penalty for a first offence; death was the penalty for a second offence!

In England William Vaughan in his Directions for Health first published in 1600 warned men about the way tobacco damaged their virility, "wasted the oil of their vital lamps" and instructed them to repeat the rhyme:

"Tobacco, that outlandish weed,
It spends the brain, and spoils the seed:
It dulls the sprite, it dims the fight,
It robs a woman of her right."

James I wrote *"A Counterblaste to Tobacco"* in which he challenged tobacco's supposed medicinal effects. Tobacco was sold mainly through apothecaries and in part James I *"counterblaste"* was provoked by a complaint from the Guild of Physicians that people were self-administering tobacco, overcoming their monopoly on health, part of their long-running battle with the Guild of Apothecaries. But already tobacco was too valuable a trade item. In 1630 2,300lbs were imported; by 1640 two million pounds were being imported. James I had tried to increase the custom duty on tobacco from 2d per pound to 6s 10d an increase of 4000 %. Naturally this was unenforceable and had to be reduced but nevertheless the income was huge. One threat to this source of revenue was the home grown cultivation of what the herbalist Nicholas Culpepper called English Tobacco or Yellow Henbane. Very importantly the economic survival of the Virginia colonies depended upon tobacco. As early as 1619 the first black slaves had been bought in Jamestown Virginia; over time tobacco cultivation became more and more reliant on slaves. The introduction by John Rolfe in 1612 of a milder variety of tobacco, more pleasant to smoke, had made Virginia tobacco more popular. In 1619 James banned the cultivation of tobacco in England to protect the Virginia trade and to preserve his customs revenue. Price protection gained another boost in 1730, by a law was passed to control the quality and quantity of tobacco grown in Virginia: tobacco had to be exported through tobacco warehouses, allowing poorer quality and "excess" tobacco to be destroyed. Meanwhile tobacco became the staple of the smuggling industry; as Kipling put it

"Five and twenty ponies trotting through the dark-
Brandy for the Parson, and 'Baccy for the Clerk ."

Nothing could stop the use of tobacco. In the Ottoman Empire in 1647 the Mufti of Constantinople faced up to the inevitable and permitted its use. Tobacco joined opium, wine and coffee as the four "cushions on the sofa of pleasure" . Turkey became a major exporter of tobacco and the home of the cigarette. Elaborate rituals arose involving the sharing of highly decorated pipes: the honors of the pipe symbolized acceptance of diplomats in the Sultan's palace. In 1700 in Russia Peter the Great and in 1724 Pope Benedict IV repealed the anti-tobacco laws; both are smokers! One wonders if peter got the habit while visiting London in 1698. The rise of cigarettes to worldwide popularity began when Philip Morris started to sell Turkish cigarettes in London in 1847, but did not really take off until the invention of a cigarette rolling machine in 1884 by James Bonsanck. Instead at first snuff and then cigars were the most favored means of using tobacco. Important in the success of cigarettes was the 1839 accidental discovery by quickly drying the leaf of "Bright" tobacco, or golden Virginia tobacco, by a slave named Stephen, and the later introduction of "White Burley" tobacco leaf.

It was 50 years until the first health concerns were raised, because of the correlation between the incidence of cancer and smoking first noticed by researchers in Cologne in 1930.

Getting the buzz: Cannabis sativa

Leaves, fruits and also the resin obtained from female plants of Cannabis are utilized to get a buzz. There is an extensive jargon associated with cannabis : hashish, marijuana, grass, pot, dope, ganja and kif are just some. And a marijuana cigarette is a reefer, joint or spliff. The resin is collected by rubbing the flower heads with the hands or a leather apron and then scraping it off and pressing it into blocks. In Nepal one story was that men ran naked through the cannabis fields to collect the resin, afterwards scraping it from their bodies, nice work if you can get it, but probably not true. Another method was to shake and sieve dried plants to collect up the resin particles. A highly concentrated hash shish oil has also been produced by extracting the active THC from the resin with solvents.

Eaten or drunk the effects of Cannabis can take an hour to have an effect, last for several hours, but are less easy to control. Smoking is much quicker in effect and more easily controlled, but it was only with the spread of tobacco smoking that cannabis smoking became widespread. Marijuana or grass, the unprocessed dried plant has less concentrated THC so that it takes a while for the buzz to take effect. Today hashish is crumbled with tobacco and rolled into joints. Commercial tobacco has additives like saltpeter to keep it alight. However initially when smoking was introduced it was mainly in pipes. Pipes with a bead in the stem to prevent hot ash being inhaled, called chillums or chuillums in India, develop. Alternatively special pipes that cooled the harsh cannabis smoke through water, hookahs, hubble-bubbles or bongs, were being used, probably first in Persia. Probably arising among American soldiers in Vietnam was the habit of breathing in vaporized hash shish oil either impregnated into tobacco or a cigarette paper or heated in a small glass pipe.

Cannabis has been utilized as a medicinal remedy for many millennia. The name has an ancient Sumerian root. Cloth made from cannabis fiber has been dated to 9,000-10,000 years ago but the first pharmacological use is recorded as such in Chinese herbals dated to the emperor Shen Nung more than 4000 years ago. In the ninth century B.C.E. Mesopotamia it was used as incense. It has been suggested that it was a component of the Egyptian incense kiphy along with scents such as myrrh and calamus but there is not much evidence for the use of hemp either as a fiber or medicinally in Egypt. The mistake perhaps was to equate kiphy with the Arabic word for Cannabis, kif. However a biblical link was made by Sula Benet in 1936 by claiming that the holy anointing oil of the Jews which contained myrrh, cinnamon, kassia and calamus actually contained cannabis not calamus. The interpretation rests on the etymology of kaneh bosm; bosm is aromatic but kaneh might mean reed or hemp. The oil was used as part of ceremonies when the priests entered a trance like state to speak to god. "Messiah" meaning "anointed one" refers to this ancient practice. It is all very intriguing if not at all certain.

What is more sure is that Cannabis was (re)introduced to the Near East and Middle East about 800 B.C.E. by the invasion of the Scythians from the mountains of the Altai in southern Siberia. Herodotus the Greek historian reports that they had the habit of purging themselves by sitting in a kind of sauna in which they threw cannabis inflorescences on hot stones. Zoroaster placed cannabis at the top of his list of medicinal herbs in his Zend-Avesta written in 550 B.C.E. Cannabis was definitely known medicinally in Roman times. In the first century C.E. it is mentioned by Dioscorides as a remedy for earache and to suppress sexual urges, a kind of bromide in the tea or saltpeter in the coffee. In the second century Galen describes how hemp cakes in moderation give a feeling of well-being but in excess intoxicate and make the eater impotent.

This classical knowledge of Cannabis survived the decline of the Roman Empire in the Middle East and the use of hashish recreationally became well established in the Arabic world. Islamic attitudes are complicated because the Koran does not prohibit intoxicants but only intoxication. Mohammed had a drunkard whipped not because he was drunk but because as a consequence he did not perform his duties. Wine remained, at the very least, an element of Arabic medicine, in which other drugs were dissolved. Arabic poets continued to praise it. Omar Khayyam praised wine.

Similarly cannabis was not prohibited. It was eaten in sweetmeats, such as majoon, made with honey and spices, or stuffed in dates or added to Turkish delight. The tales of the Arabian Nights have several hashish eaters, normally as figures of fun. Or it was drunk in a tea. Bhang is the Indian variety drunk today made with milk to dissolve the cannabis resin. The slight ambiguity about whether Cannabis was actually included as an intoxicant, and its medicinal use, allowed its users to evade proscription and it remained part of Arabic life for many, especially the poor. In particular the mystical sect of Sufi Muslims that arose in the late seventh century B.C.E., used hashish to obtain a state of spiritual ecstasy. The Sufis were denigrated by mainstream Islam and hashishiyya (users of hashish) became a derogatory term. Several attempts were made to destroy the Sufi sect and in the process Cannabis use became targeted. It symbolized the religious and class challenges of Sufism to the orthodoxies of the rulers of the time.

Later this term was thrown at another schismatic sect the Nizari Ismaili, founded by Hasn ibn-Sabah, a friend of Omar Khayyám about 1090. The ascetic sect promoted their cause like a latter day terrorist group, by a mixture of political murder and proselytizing orthodox Islam. From this came the myth of the hashshashin, assassins who lived by robbing passing trade caravans and used a cannabis beverage to induce a state of bliss. The story was first embroidered by Marco Polo that initiates to the sect were first drugged before being allowed into a beautiful garden paradise where fountains ran with milk and honey and beautiful houris satisfied their every need, as a vision of the paradise they would enter if they died while acting for their cause. The suicide bombers of today are their spiritual descendants but in the meantime Marco Polos myth became one of the pillars of the western orientalist view of the East.

In the west perhaps surprisingly Cannabis was not particularly valued except as a source of fiber. It was not brought back by crusaders neither did it become a fashion when the fashion for drinking coffee did. the expansion of the trade empires of the Atlantic coastal nations did bring reports come back about it but it was only in the late 18th and early 19th century, along the orientalist fashion of the time and the exposure of western colonialists to eastern habits, did it emerge as a significant recreational drug in the west.

The joy-plant Papaver sominiferum

There is no better example than the history of early human attitudes to the special plants than the history of opium . Opium now has a terrible reputation but it was not always like this. The earliest record of it is Sumerian where its ideogram is represented as "joy-plant". In Greek myth the juice of the poppy was used by Demeter to soothe the grief of rape and the loss of Persephone to Hades. Helen of Troy gave "*nepenthes*" to Telemachus and his band to help them forget their brother warriors who had been killed. Theophrastus in the 3rd century B.C.E. referred to the latex obtained from poppy capsules as opium and meconium the juice obtained by crushing the plant . As well as being a narcotic morphine is an analgesic and hypnotic. Among the many other alkaloids that opium contains there are several others with an established therapeutic effect. Noscapine, codeine and papaverine are antitussives, suppressing coughs, and analgesic and pain. Papaverine also acts as smooth muscle relaxant. Noscapine and papaverine are not a narcotic but codeine is a mild narcotic.

According to Pliny the poppy *Papaver* gets its name because poppy juice was used in baby food to help them sleep; baby food is papa, so true baby food is Papaver, and indeed in the Ebers papyrus of 1500 B.C.E. opium is a component of a concoction used to sedate children. The Assyrians invented the method of scratching the poppy capsules to get the latex, a method that was rediscovered and described in 40 C.E. by Scribonius, physician of the Emperor Claudius, in his *Compositiones Medicamentorum*. In Roman times opium was used as a panacea. Dioscorides wrote

"It completely takes pain away. Ameliorates coughing, stops stomach fluxes, and is given to those who cannot sleep."

But he also warns that *"being drunk too much   ...it kills."* Dioscorides described how to pound poppy heads with leaves in a mortar and press the paste into pills. Dioscorides described how the latex was stronger than the juice of the crushed plants.

It was piece of latex about the size of a bean rolled with honey and dissolved in warm wine that Marcus Aurelius, the stoic emperor-philosopher, started every morning. Mithridatium was one name for this panacea, named for the famous antidote to poisons devised for Mithradates king of Pontus in Asia Minor in the second century B.C.E., an ancient treacle, or *theriac*, antidote to poisons. It seems likely that much of his magnificent acceptance of the ways of the world, sometimes feeling like deep world-weariness, was due to his opium breakfast? He seems to have been able to manage his habit; Claudius Galen (c.129-216 C.E.), Marcus Aurelius' doctor, and the most famous physician of the Roman Empire after Dioscorides, reports that Marcus Aurelius reduced his consumption when it was necessary to fulfill his imperial duties .

Opium remained an important part of repertoire of physicians trained on Dioscorides or Galen's works. It was used in all sorts of tinctures and theriacs or treacles, whose ingredients were part of the secret lore of physicians and herbalists. Those devised in Cairo and

Venice became famous. Mithridatium was still in use into the 18th century as a general panacea. Philonium Romanum was another other famous opium based remedy named after its reputed inventor, Philon a first-century B.C.E. physician of Tarsus. Diascordium included a range of other herbs such as *Teucrium scordonia* water germander, valued as an antidote for poisons and as an anti-venereal disease. Paregoric tincture, or camphorated tincture of opium, still provides a remedy for diarrhea. The most famous remedy of all, and one that came to be used recreationally, was *laudanum*.

Paracelsus, the pseudonym of Theophrastus Bombast von Hohenheim, a German physician of the early 16th century, is reputed to have first used the term *laudanum*. His laudanum was a mixture of one quarter opium, along with henbane, crushed pearls, coral and amber, as well as musk and stag's heart, bezoar stone from cow's intestines, and even unicorn! In the 1660s George Sydenham popularized the use of the term laudanum for a solution of opium in alcohol. Sydenham's recipe was two ounces of opium, one ounce of saffron, and a drachm (an eighth of an ounce) of cinnamon and cloves powder, all dissolved in a pint of Canary or sherry wine in a steam bath for 2-3 days. Sydenham supplied laudanum to both Oliver Cromwell and King Charles II. Charles is reputed to have said

"Among the remedies which the almighty saw fit to reveal to man to lighten his sufferings, none other is universal and effective."

The use of opium recreationally became more widespread as it gained reputation as an aphrodisiac; its effect was to delay male climax and so it seemed to increase male ardor. More importantly as Richard Davenport-Hines has so eloquently pointed out, from the 17th century onwards as more and more people turned away from the certainties of religion and to a consideration of the self, so they also turned to drugs to feed their souls , as an aid to contemplation and expansion of consciousness. The use of laudanum and other tonics grew as people became more and more alienated from a society becoming more industrialized and competitive. As well as laudanum there were many other patent medicines available over the counter. Thomas Dover, a pupil of Sydenham invented one famous one called Dover's Powder. It was a mixture of opium and ipecacunha (*Psychotria ipecacuanha*) from Brazil (that contains alkaloids about 70% emetine, 25% cephaelin and a trace of psychotrine) to induce sweats. The addictive effects of such opium based drugs were not fully understood.

Nineteenth century experimentation and commercialization.

The drug cultures of the 19th century are particularly relevant to present-day attitudes to recreational drugs. As always attitudes varied over time and place and in social class. Opium, especially as laudanum, remained the main palliative of the middle and upper classes, and was a luxury used only medicinally by the working classes. Its use was widespread, though disparaged. The unstated counterpart of Karl Marx's quote that "religion is the opium of the masses" is that in the nineteenth century is that opium was the opium of the middle and upper classes. The Prince of Wales, later George IV, was an addict who binged on cherry brand and laudanum. Coleridge was called by Wordsworth the "slave of stimulants". Coleridge tried repeatedly to reduce his consumption of opium but lied to himself. After 1820 he lived in the house of his physician John Gillman to manage his addiction but made clandestine visits to a local druggist to get extra supplies of laudanum. Elizabeth Barret was an inveterate user too and Prime Minister Gladstone's sister was an addict. Mrs. Beeton in her book of "Household Management" recommends laudanum for several illnesses including the common cold. But the use of opium based products was widespread not just among the middle and upper classes: the poor turned to patent medicines or "physic" various medicinal remedies, such as "Godfrey's Cordial" bought over the counter, if not opium itself. In Elizabeth Gaskell's "Mary Barton", John Barton is clearly an opium addict in novels Many women became addicts as a result of the use of laudanum to ease the pains of child birth and after. Paregoric was a favorite to keep teething babies quiet.

The recreational use of opium and later morphine was established only among a relative few in the west. Thomas De Quincey's *"Confessions of an Opium Eater"* described eloquently both the pains and pleasures of taking opium, De Quincey had the habit to take opium before going to the opera. Opium increased his pleasures of his walks through the streets and markets of London on a Saturday night: opium relaxed him opened him to new experience, he joined in with the lower classes, he explored the bye-ways as an explore would terra incognita. In Charles Dickens' Oliver Twist the artful Dodger and his gang sit round a table to smoke log clay pipes that probably contained a mixture of opium and tobacco.

Not just amongst the intellectual elite, and the upper classes was "scientific" experimentation was valued. Chemists like Joseph Priestley were discovering and synthesizing new compounds, and half-killing themselves in the process by sniffing their products. Meanwhile scientific showmen like Humphrey Davy and Michael Faraday were turning their discoveries into entertainment. Huge numbers of people attended scientific demonstrations at the Royal Institute. The physiological effects of nitrous oxide, laughing-gas, one of Joseph Priestley's discoveries, was first demonstrated by Humphry Davy. It was a great crowd pleaser. Others turned to the drugs of the east.

It was the growing interest in the East that gained a tremendous boost with the Napoleonic adventure in Egypt at the end of the 18th century that encouraged the use of cannabis in Europe. Soldiers had become accustomed to it in the alcohol free Europe and brought the habit back with them. Napoleon tried to ban its use because of its effects on discipline but three scientists on the expedition Silvestre de Sacy, Rouyer, and Desgenettes experimented with it and brought the habit back. Scholars were fascinated by the history and culture of the east, and among the intellectual elite of France it use was thought to inspire the imagination. There was a fad for everything oriental and that included the drugs. A French translation by the Viennese Joseph von Hammer-Purgstall provided the first full story of the hashshashin. Many authors were to use the story. The Arabian Nights were translated. A French doctor called Jacques-Joseph Morceau used it to treat to calm his patients and used it himself. He supplied his intellectual friends including the novelist Théophile Gautier and through him the *Les Club des Hashshashins* was founded in Paris about 1844. There were monthly meetings where the members, that at various times included many luminaries of the French intellectual scene, Gérard de Nerval, Charles Baudelaire, Victor Hugo and Honoré de Balzac, Flaubert perhaps even Eugène

Delacroix and Alexandre Dumas, could enjoy the hallucinogenic effects of the resin in convivial surroundings. In the Count of Monte Cristo there is a retelling of the myth of the hashshashin.

They ate dawamesk, a green hashish jam, made with almond and pistachio nut paste, orange and tamarind peel, cloves and sugar. By eating it they passed through all the stages of intoxication: hilarity, relaxation, stupor, hallucination. Hashish was not the only drug participated in. Baudelaire was an opium addicted alcoholic, indeed he rarely participated in hashish and in his very influential description of the effects of hashish he conflated the effects of opium and cannabis.

Because Coca leaves lose their power rapidly after being harvested their use did not transfer to other parts of the world until the 19th century. At first cocaine became a component of tonic drinks. One favorite was *Vin Mariani*, was recommended by everyone according to its inventor the Corsican Angelo Mariani, from Queen Victoria to Thomas Edison and Émile Zola, and three different Popes. Pope Leo XIII even presented Mariani with a gold medal "in recognition of benefits received form the use of Mariani's tonic". In 1885 the American John Styth Pemberton launched "French Wine Coca" in competition but his non-alcoholic version with added Kola and sugar – Coca Cola was to become such a success that it came to symbolize America. The level of cocaine in Coca Cola was always very small because cocaine soon breaks down in dead coca leaves. However, towards the end of the century paranoia about *"drug-crazed niggers"*, such as an article linking cocaine to black violent crime published by the New York Tribune, resulted in Coca Cola playing down its "medicinal" benefits and instead emphasizing it's refreshing nature, in other word, more its caffeine than its cocaine content.

But for a while in the late 19th century there were many drug-based tonics available. The next generation of French poets, writers and artists, the inheritors of the tradition of Baudelaire, such as Verlaine and Rimbaud had a wide range of drugs available. Many artists favored the green fairy, absinthe, so memorably recorded by Degas' painting of L'absinthe; the pretty but somewhat towselled young woman, like a fading flower, deep in her own thoughts with a raffish bearded pipe-smoking flaneur beside her. There is another earlier painting of an absinthe drinker, but this is by Edouard Manet and shows a much more romantic view, of a handsome man in a tall hat swathed in a dark cloak.

Meanwhile the development of the chemical sciences led to the purification of the active compounds in natural plant drugs. A Parisian pharmacist Pierre-Jean Robiquet discovered the alkaloids nicotine and caffeine and perfected the techniques for the extraction of morphine in 1806. By 1825 the pharmacist Heinrich Emanuel Merck had established the eponymous drug giant by the large scale manufacture of morphine but he was not the only one; Thomas Morson in London, McFarlane & Company in Edinburgh were also manufacturing it. Codeine was produced in 1832, cocaine in 1860, heroin in 1883 and mescaline in 1896. The opiates first morphine and then heroin (diacetyl morphine) were marketed as replacements for opium. They were pure and stronger so that much less was needed to gain the same effect. Morphine was a great boon and was widely used in 19th century conflicts to ease the suffering of the wounded. The immediacy of the effects, and seemingly also the control of its effects, were increased by the invention of hypodermic (meaning below the skin) injection. Florence Nightingale was one of the advocates of morphine injection. Not just heroin was injected. Sherlock Holmes injects a 7% solution of cocaine. At first heroin was thought not to have the negative consequences of opium/morphine dependency but by the end of the century addiction to it had been noted. Hence when heroin was developed by Bayer and given the name *heroisch* "heroic drug', it was five times stronger than morphine, it seemed a way of escaping the negative effects of morphine. It was marketed by Bayer along with their other panacea Aspirin. As heroin seemed at first to be a solution of a morphine problem so methadone seemed to provide a powerful analgesic without the dangers of addiction. A series of synthetic narcotics were also marketed in this way.

In 1860, Albert Nieman isolated cocaine hydrochloride from coca leaves by a process that in a modified form is used to=day by the cocaine manufacturing industry. First cocaine sulfate, also called pasta, basuco, basa, pitillo, paste is extracted by crushing the leaves, for example by treading on them, in a dilute solution of sulfuric acid. The coca sulfate paste that is produced can be mixed with tobacco and smoked. Chemically stable cocaine hydrochloride is produced by dissolving coca paste in a mixture of hydrochloric acid and water, and adding potassium chloride to remove undesirable substances, and then ammonia to precipitate the cocaine powder. Alternatively, the paste is by washed in a series of organic solvents, first kerosene which is then so that crystals form, then the crystals are re-dissolved in alcohol and re-crystallized by the addition of sulfuric acid. Further washing takes place with potassium permanganate, benzoyl and sodium carbonate.

In 1884 Freud promoted the use of cocaine as a means of controlling morphine addiction and cocaine. between 1884-1887 Freud wrote five papers on the value of cocaine, but he later realized his addiction and called cocaine the "Third scourge of Mankind" (the first two being alcohol and heroin).

Both in Europe and America drug use/abuse/addiction came to be seen as a foreign vice. Opium was identified with the Chinese. Cannabis with Mexicans and blacks. Chinatowns in London's East End or San Francisco became notorious for their opium dens those these were more fictional than real. In one episode Sherlock Holmes visits one.

There was, what seems to us, an astonishing double standard; drug use was alright, the drug trade promoted as an essential part of trade, as long as it was only foreigners who paid the price . Trade in opium helped to pay for tea. The East India Company had a monopoly on the trade of opium in India and they exported 1500 tons of opium each year worth a billion dollars in today's values. One ton of opium could pay for nearly 40 tons of tea. As a result from the 18th century onwards opium was much more widely available in China. Not just the British were involved in the trade. American traders bought their opium, which was of higher quality from Istanbul. The Chinese government made several attempts to limit this dangerous foreign import. Its import was banned both in 1729 and 1800 but the ban was widely flouted by the western trading nations. In 1839 the Chinese government's confiscation and burning of a year's supply precipitated the first opium war between Great Britain and China. The result was that Britain was ceded Hong Kong and granted favored trading status, along with France and USA. Trade conflict continued and after a second opium war in 1856 France and Britain were given further rights. Chinese culture shielded for millennia was now exposed to western influences, including Christian proselytizing. In addition secrets of the tea industry were revealed.

The War on Drugs

At the beginning of the 20th century many patterns of modern recreational drug use had become established.

A strong puritan strain against alcohol and narcotics in the USA has had a malign influence both at home and abroad. The criminalization of the drinking of alcohol and the use of narcotics has only made new criminalized large sections of the population, led to the expansion of criminal activity, and signally failed to reduce the number of alcoholics or drug addicts. The Pure Food and Drug Act in the USA effectively ended the sale of narcotics over the counter in the USA. The Harrison Act of 1914 criminalized drug addicts and restricted opiates and other narcotics to the prescribed treatment of disease. Police officers posed as addicts an entrapped doctors. It has been estimated that 30,000 physicians were arrested between 1914 and 1938; 3,000 served prison sentences for supplying addicts . Pharmacists were also targeted by the police. Another consequence of the Act was the movement of drug users to the cities where an illicit trade arose to supply their needs. In the process drug use patterns, previously opium and morphine use had been a largely middle-aged and middle class addiction, expanded into the lower classes and there was a shift to heroin and cocaine abuse.

In 1919 an amendment to the Constitution and the Volstead Act brought in the prohibition of alcohol. Both the drug user and the alcohol drinker were criminalized and so inevitably criminal empires arose to supply them. A further notch on the ratchet of criminalization came with the foundation of the Federal Bureau of Narcotics in 1930. It successfully conflated the drugs with criminality. Repeal of the Volstead Act in 1933 by adoption of the 21st Amendment brought only partial respite. The criminal empires just moved into drug trafficking. And by now the medical profession had been co-opted: "drug addiction" was first listed in the American Psychiatric Association's diagnostic handbook as a disease.

Marijuana was the next easy target. By the late 19th century hashish smoking parlors were found in every American city with 500 New York City. However the tide had turned, perhaps partly because, following the Spanish-American war, at a time when more and more Mexicans were entering the country as farm laborers they were caricatured in Randolph Hearst's newspapers as pot-smoking layabouts. In 1914 the US congress passes the Harrison Narcotics Act, to control recreational use of drugs. Hearst's tabloid campaign continued with articles having such headlines such as "killer weed from Mexico" and "Marihuana Makes Fiends of Boys in 30 Days" and telling tales of "marijuana-crazed negroes" raping white women. In the cities homosexuals were reputed to be users. In 1937 the Marijuana Tax Act imposed huge taxes and restrictions, with accompanying penalties, effectively criminalized the use of Cannabis. The use of tax law enabled federal control to be exerted. Huge swathes of the American people were criminalized. By the 1980s 300,000 Americans were being arrested annually for offences related to cannabis use.

After the second world war with the loss of the Nazis and the Communists as public enemies the source of societies ill, drug abuse was identified as Public Enemy No. President Nixon launched a war on drugs in 1971 but already the war on drugs had been internationalized. First the concerns of American missionaries in the Philippines and china led to the meeting of the International Opium Commission in Shanghai in 1909. This led to the Hague Convention of 1912 in which the signatories agreed to "control the preparation and distribution of opium, morphine, and cocaine.". In the same year the Hague convention was incorporated in the Treaty of Versailles. Over the next decades an international system for the control of narcotic drugs has come into being. Three treaties under the aegis of the United Nations now seek to control the international trade in narcotic the 1961 Single Convention on Narcotic Drugs which established the International Narcotics Control Board, INCB; the 1971 Convention on Psychotropic Substances; and the 1988 United Nations Convention against Illicit Traffic in Narcotic Drugs and Psychotropic Substances.

In parallel American agencies has repeatedly sought to extend the war on drugs outside the borders of the USA, and has at time become corrupted in the process, using the drug trade as an armament in the aggressive pursuit of foreign policy aims.

Cocaine and cannabis have become the modern drugs of choice along with a whole load of synthetic drugs from amphetamines to ecstasy. Cocaine snorting first became popular in the USA about 1905. The mark up from the unprocessed leaves to the powder sold on the street is enormous, perhaps more than 1500 times. Not since nutmeg has such profits been available. Little wonder then that the cocaine trade nearly subverted the government of Columbia. And then there came crack cocaine that can be smoked.

The first British legislation to control over-the-counter medicines was passed in 1860 but in the UK throughout there has generally been a pragmatic approach emphasizing treatment rather than punishment of drug users. Although the level of substance misuse has fluctuated in the UK over the centuries, it reached a low in the early 20th century .

A scientific survey of users of recreational drugs in the US indicates that they are well educated, employed full-time, participate in recreational and community activities, and have physical and mental health as good as the general population . But at least for some people the plants have their own trick, the chemicals they produce are so often addictive. We can chase the dragon, or as the hippie group the Jefferson Airplane sang, we can ride the tiger, but it is very difficult to dismount without being bitten.

Phantastica: opening the doors of perception

It was partly through his experience of mescaline from Peyote (*Lophophora williamsii*), the sacred "mushroom" of the Aztecs, actually a cactus, that Aldous Huxley wrote about drugs opening the doors of perception. There are several other cacti that have been utilized by native peoples for their psychoactive effect. The dried peyote is sliced into buttons for consumption. It is rich in alkaloids; the principle one (30% of total alkaloid content) is mescaline. Three hours after consumption visual, auditory, olfactory and tactile hallucinations begin; they last for three days.

The phantastica, the psychedelic or hallucinogenic drugs are in terms of their social reach of relatively little significance, far exceeding the passionate concerns about the "abuse" of drugs, and the scare stories they have initiated. And yet in many ways these drugs are closest to a traditional use of plants in religion and magic.

Many of the active chemical components of these plants are alkaloids. They are considered physiologically safe, and also said by some not to be addictive, but whether this is because of the profundity of their effect that they are generally only used occasionally it is difficult to say. They can have a profound and lasting effect on the psychological well-being of the user.

Hallucinogenic drugs have always played an important part in human spirituality and have normally been administered in a highly ritualistic manner. The Priestess of the Oracle at Delphi uttered her prophesies while drunk on the vapors, that issued from a cleft in the rocks beneath her feet, likely from some burning plant material. Escohotado has made the point clearly that inebriation was sometimes religious in nature . He has distinguished two types of sacramental inebriation: possession inebriation and journey inebriation. Possession inebriation induces "raptures of bodily frenzy where critical consciousness disappears" and rests upon drugs like alcohol. Journey inebriation is less often accompanied by music and dancing, a party atmosphere and is more likely to be shaman led. It relies on hallucinogens that empower the senses. However several drug plants transcend this Cannabis though normally a possession inebriant can induce hallucinations. "Speedball" a potent mixture of heroin and cocaine used by some intravenous users interacts with the μ-opiate and D1-dopamine receptors of the *nucleus acumbens*.

In different parts of the world one or several plants have been used to induce a hallucinogenic experience. In many different places this has involved, as well as the use of plants, the use of "magic" mushrooms. In Europe the history of hallucinogenic inebriation is rather obscure. Two millennia of Christianity that prefers and elevates the use of fasting to obtain an altered state has pushed traditional folk hallucinogens to the margins, demonizing them and identifying them with witchcraft. Valentinian, the Roman Emperor between 364–375 C.E., applied the death penalty for nocturnal religious ceremonies or magic, thereby bringing to an end the *Rites of Eleusis*. In 743 C.E. the Merovingian king Childeric III condemned all pagan magical practices.

The establishment of the Inquisition brought out from time to time "evidence" of witches who rubbed ointment into their armpits and vagina, sometimes with the use of a broomstick or walking stick as a kind of dildo, to enable them to "travel" . There are several artistic representations such as a painting by Franz Francken that has been called *The Witches' Kitchen* shows several women at various stages of undress, and all the paraphernalia of making witches potions; several pornographic images, by Hans Baldung Grien, a pupil of Durer, show similar scenes, indeed one called *The Witches Departing for Sabbath* shows naked women in trance-like states, one of whom is either masturbating or applying unguent to her fanny. These are highly misogynist representations and it is very difficult to disentangle these accounts, all by men, from the context where women's sexuality was controlled by men, and when not controlled it was feared.

But what went into these witches potions, if they existed. Shakespeare's witches in Macbeth include hemlock and yew but very likely their hallucinogenic ointment would have included plants rich in tropane alkaloids such as Deadly Nightshade (*Atropa belladonna*), and Henbane (*Hyocyamus niger* and *H. albus*). The former was used as a cosmetic anyway to dilate the pupils. The discovery of the hallucinogenic properties of *Hyoscyamus albus* was attributed by the Ancient Greeks to Hercules   The thorn-apples (*Datura stramonium, D. arborea* and other species) and angel's trumpets or floripondio (*Brugmansia*), and chrisanango (*Brunfelsia*) are related plants, and another rich source of tropane alkaloids. They have been used widely in the America as hallucinogens. They are used by shamans to maintain a relationship with the sangariite, the invisible or pure ones of the forest from whence come new cultivars of their crops . In Peru shamanistic practice maintains a connection called *vegetalismo* to traditional knowledge about the use of plants The main active component, hyoscyamine, is a stimulant of the central nervous system, acting on cells connected by the post-ganglionic cholinergic nerves to the parasympathetic system .

Around the world the range of other plants used to induce a trance-like state is very wide. In Africa the iboga vine *Tabernanthe iboga* (Apocyanaceae) is used in the initiation rites of several cults. In one cult the hallucinating initiate sees the god plant Bwiti. Shamans use iboga to contact ancestors and the spirit world, but it is also used traditionally to prevent hunger and fatigue and as a general tonic. Ibogaine is the main alkaloid present. Recently it has been controversially suggested that it may provide a therapy for addictions, including to alcohol, cocaine and opiates , though one wonders if this is not just another case of a new drug replacing an old one as morphine replaced opium and was replaced by heroin and cocaine. In Africa many other hallucinogens plants have been used traditionally such as Kwashi *Pancratium trinathum* by bushmen in Botswana and Kanna (*Mesembryanthemum expansum* and *M. tortuosum*) by the Hottentot.

The Americas are particularly rich in hallucinogenic plants. Ebena snuff, is used in shamanistic rituals by Amazonian natives. It is made from three main plants: *Virola theiodora* (Myristicaceae), *Justicia pectoralis* (Acanthaceae) and *Elizabetha princeps* (Leguminosae). It is the *Virola* species that contains the rich mixture of psychoactive tryptamines and beta-carbolamines. The other species, E. princeps is used as ash, and J. pectoralis calcium carbonate crystals, aid the extraction of these psychoactive components. Similar compounds are found in the legume, *Anadenanthera peregrina* (niopo or yopo). Archaeological excavations of the remains of the Chavin culture of Peru dating from about 400 B.C.E. ceremonial ingestion of hallucinogens by the discovery of objects such as small mortars to grind, and bone tubes and spatulas to snort *A. colubrine,* as a sort of hallucinogenic snuff called vilca.

Another South American hallucinogenic plant is called ayahuasca, the "vine of the soul", *Banisteriopsis caapi* (Malpighiaceae) and *B. inebrians*. Different varieties of the former give visions of different colors and content. The active chemicals are the beta-carbolines harmine, harmaline, tetrahydroharmine but the composition and concentration varies geographically. Some Amazonian tribes, including the Yahua, use the vine alone but normally the vine is combined with another plant, one containing tryptamine alkaloids, especially DMT (demethyltryptamine), usually from chucruna (*Psychotria viridis* or other *Psychotria* species) or oco yagé (*Diplopterys cabreana*) a close relative of the Ayahuasca vine. DMT is also found in *Mimosa hostilis* (*M. tenuiflora*) which the Pancaru Indians of Pernambuco in north-eastern Brazil prepare the hallucinogenic drink *vinho de Jurumena* and it is in the seeds of *Piptadenia peregrine* (along with bufotenine) from which Indians of the Orinoco Basin and Trinidad prepare a snuff called yopo. DMT is found widely in other plants.

Many hallucinogens appear to have an effect by enhancing serotonin (5-HT) activity in the brain by stimulating 5-HT2A receptors, especially those expressed on neocortical pyramidal cells, increasing pre-frontal cortical metabolism and altered states of consciousness. Mescaline, like many of the other hallucinogenic compounds, N,N-dimethyltryptamine from in ebena snuff, harmine from the ayahuasca vine, mimics the chemical structure of the brain messenger compound serotonin. At high and nearly toxic concentrations hallucinogenic visions are produced but at lower concentrations they bring tranquility and peacefulness. The Ayahuasca effect is actually rather complicated because the beta-carbolines inhibit the neuronal enzyme monoamine oxidase so that DMT, mainly from the companion *Psychotria* plant, is active when ingested orally. DMT is a substance secreted by the human pineal gland and is highly hallucinogenic when smoked though it is in relatively low concentration in *Psychotria*. A different and more lasting experience is produced if oco yagé is used instead. The "Ayahuasca" experience is clearly greater than that which either of its components elicits alone.

The ladder of delusion and self-deception

The dangers of abusing hallucinogenic plants are aptly illustrated by the abuse of *Salvia divinorum* from Oaxaca in Mexico that has become in recent years a fad. Known as *Yerba de María* ("Herb of María"), *hojas de la Pastora* ("leaves of the Sheperdess"), *Hierba de la Virgen* ("Herb of the Virgin") it contains a compound Salvinorin which is said to be the most potent naturally occurring hallucinogen so far isolated. When the herb *Salvia divinorum* is consumed by chewing the fresh leaves, the effects are usually (but not always) pleasant and interesting because the amount absorbed is very small, but when the dried leaf is smoked, the hallucinogen is vaporized the consequences may include a complete loss of awareness of, and control over, the body. In this state the user can damage themselves physically without being aware of it.

Probably of more social significance is the increasing evidence of a relationship between a hallucinogenic trip and a subsequent psychosis such as schizophrenia. One is exogenously chemically induced and the other endogenously. One is of normally short duration and the other episodic but of much longer duration. It is becoming increasingly obvious that for some people, such as regular cannabis users, the one can lead to the other, as the brain tries to adapt to the chemical challenge that the recreational drugs provide.

Plants provide a great treasure, rich in potential benefits for humans but the use of plant chemistry recreationally is fraught with dangers. The human brain is a highly adapted and adaptable system and to bludgeon it into unusual kinds of activity with exogenous chemicals, like feeding a car engine with a too rich fuel mixture it will not function properly but instead it will race or stall, or it will blow up.

The influence of hallucinogenic drugs on a few artists, writers and musicians has been profound. Rock music has at times seemed synonymous with the abuse of hallucinogenic and other recreational drugs. Aldous Huxley's famous phrase about drugs opening the doors of perception, was taken up as their name by the rock group the Doors. William Burroughs and Philip K. Dick are two writers that have produced work of startling vision. The gonzo journalist Hunter S. Thompson appears to have been out of his head a large part of the time. For these people drugs appear to have enhanced their creativity. However rather too often, far from inspiring great art, drug use has only resulted in a withdrawal from artistic endeavor and a desperately sad personal decline. Far from opening the doors to perception drug users have only climbed a ladder of delusion and disintegration.

The modern world provides so many other ways of getting kicks, to stimulate the brain though the arts and sport There is a deep-seated and ancient ambivalence about the use of plants for the soul rather than for anything other than the primary ones, of food or craft, an ambivalence that has included at times not just the extraordinary hallucinogens but also the medicinal and the mundane such as coffee or tobacco. There is less ambivalence about the value of plants for their beauty, and the way they link us to nature, to assuage the trials of the soul, but that is part of the subject of the next chapter.

8 Return To Eden

There are two strands to the cultural history of plants: firstly the story of the human economic exploitation of plants that has enabled us to grow to dominate the world, and secondly, but running in parallel, the story of our imagining of plants and their meaning to us. These two strands, material possession and psychological meaning, seem superficially to be quite different and yet they are closely related and one has informed the other. They are plaited together most intimately in the story of gardens and gardeners, and in the response of writers and artists to nature, especially wild and tropical nature.

Knowledge is power

Botanical knowledge has always been a source of power. Primitive humanity must have had a wide knowledge of the plants which provided food or provided remedies for illness. By trial and error they knew which plants were poisonous and which tasty. There is very little evidence recording this early knowledge but in surviving simpler cultures closest to our hunter-gatherer forbears those with the knowledge, the shamans, the wise-women are held in respect. Knowledge of the use of plants as a source of drugs was mystical, magical and powerful.

On Egyptian papyri from 3 500 years ago there are the first lists of medicinal plants. The power was jealously protected – different kinds of plants were given secret sacred names *Ambrosia maritima*, a ragweed, now used to flavor liqueurs, was called vulture's heart.

The first recorded plant hunting expedition took place about 1500 B.C.E. when Queen Hatshepsut of Egypt dispatched five ships to gather valuable plants, animals and precious goods from the Land of Punt, a trip instructed by the Oracle of the god Amun. Punt was the personal pleasure garden of Amun and was the source of myrrh (Commiphora) and frankincense (Boswellia). Punt was probably located on the East Africa coast; Eritrea and Somalia have both been suggested. What made this expedition unusual is that it not only brought back the goods, and it brought back many of these, but it also brought back living plants. Tomb paintings show the laden ships with the plants in pots. So myrrh trees were planted in front of Hatshepsut's tomb.

Imperial expansion has often been associated with botanical discovery. Through the Alexander the Great's expedition Theophrastus (c. 372-287 B.C.E.) became aware of exotic plants. It is Theophrastus' works, dating from about 300 B.C.E., *De historia plantarum* ("A History of Plants") and *De causis plantarum* ("About the Causes of Vegetable Growth") that are the earliest surviving botanical texts, about 550 different plant species are mentioned. Theophrastus established a botanical garden in Athens too. The "History" consists of nine books: 1. The anatomy of plants; 2-5. Woody plants, 2. Domesticated plants and their cultivation; 3. Wild plants; 4. Foreign trees and shrubs, their lifespan and diseases; 5. Characteristics of trees; 6. Herbaceous perennial plants; 7. Vegetables; 8. Cereals; 9. Saps and medicine. The "Causes" has six books: 1. The growth of plants; 2. The relation of plant growth to the environment; 3. Horticulture; 4. Origin and propagation of cereals; 5. Disease and death in plants; 6. The taste and smell of plants. This represents the first scientific exploration of plants. Many of Theophrastus' ideas are probably derived from Aristotle who was his teacher as well as being the teacher of Alexander . Theophrastus was born on Lesbos where Aristotle had botanized as a young man. He took over the Peripatetic school of the "History" and "Causes" were the notes of some of his hugely popular lectures he presented for 35 years to students at the Lyceum. At one time he had more than 2000 students.

Pliny the Elder (23-79 C.E.) did not have the poetic skill of many other Roman authors but made up for it by his irrepressible curiosity. The 37 volumes of his Historia Naturalis included a myriad bits of geographical, zoological and botanical fact and fiction. There are fascinating entries about the source of cinnamon and cardamom. Pliny thought they came from Arabia and Medea, though in fact they came south east Asia, and only traveled through these lands on their way to Rome. Pliny describes many flowers and their medicinal uses but he also describes the behavior of plants, like which flower first in the spring, and which provide perfume. His Natural History was the treasure-house of knowledge for many centuries to come. Pliny was curious about everything, a characteristic which eventually killed him as he sailed into the shadow of the erupting Vesuvius to have a closer look and was suffocated by poisonous fumes.

Theophrastus makes fun of the superstitions of some herb gatherers but the next botanist of note from the classical era was the herbalist and doctor Dioscorides. Widely read, and widely travelled, probably while serving as a doctor in the armies of the Emperor Nero, Dioscorides was able to incorporate about 600 plants in his herbal, Materia Medica, produced in about 60C.E. . The Materia Medica was the foundation of botanical knowledge for a millennium and a half even though its descriptions of plants were poor. It included nearly 600 plants. Luckily a beautifully illustrated version of the Materia Medica called the Codex Vindobonensis has survived. Produced in 512 C.E. for Juliana Anicia the daughter of Flavius Anicius Olybrius the Emperor in the West, the codex has a checquered history. It first turns up in 1406 in a monastery in Constantinople. After 1453 with the conquest of the city, it was in the hands of the Turks. The Jewish doctor of Suleiman the Magnificent seems to have purloined it. Busbecq, that adventurous ambassador of the Holy Roman Emperor, who will return to our story later, saw it and managed to get some drawings for Mattioli. Seven years later a sale had been negotiated and the Codex arrived in the Imperial library in Vienna. The importance of the Codex is that there is good reason to believe that many of the drawings, which are very naturalistic in style, are derived from earlier ones drawn from nature, perhaps even by an earlier herbalist called Crateus, doctor of King Mithradates of Pontus (c.131 BC-63 BC).

Codex Vindobensis represents a peak of botanical knowledge and observation. For the millennium after it was produced there was a sad decline in the quality of copies of Herbals and scarcely any new observations were made. The collapse of the Roman Empire almost extinguished botanical knowledge in the west. Only Christian monks kept a flicker of the classical expertise alive. One remarkable exception from the generally poor quality of medieval Herbals was the De Vegetabilibus of Albertus. He was ironically nicknamed Albertus the Great because he was very short. The Pope once asked him to stand up and not weary his knees in an audience until he realized that Albertus was already standing! As a Dominican friar and Papal envoy Albertus travelled widely across Europe botanizing on his way. Once he was sent to Poland to stop the Poles killing and, sometimes eating their unwanted children. In parts of De Vegetabilibus, which was written between 1250 and 1260 Albertus records his own acute observations of plant structure. This marks him out from all other herbalists and encyclopaedists of the day, who in the main merely repeated or collected others work in a more garbled form.

The establishment of printing in Europe led to the production in printed form of several herbals that had a much older genesis. Such *incunabula* (= printed books produced before 1501), included the *"Herbarium"* of Apuleius Platonicus, probably written in Greek in the 5th century C.E. and derived from Dioscorides and Pliny, but printed around 1481 in Rome. There was also the Latin and German versions of *Herbarius* and the *Hortus (Ortus) sanitatis* printed in Mainz between 1484 and 1491. These works were little more than collections of the names of herbs. In time, the words *herbarium* became a collection of dried pressed plants, and *hortus* the botanical or physic garden, two other kinds of collections.

Superior eastern and Islamic traditions

The development of botany in China ran in parallel to that in the west. For over a millennium in the Middle Ages Chinese botany far surpassed western botany. There was a strong herbal tradition encouraged by bureaucrats and emperors. Over the centuries many different kinds of plants were included in Herbals. The earliest monograph on chrysanthemums dates from the beginning of the 12th century C.E. lists 25 different cultivars. In contrast to the west, the tradition was for a high degree of accuracy in the description and illustration of plants.

Running in parallel was a scientific study of plants in India, Vrksayurveda. In a surviving, probably pre-Buddhist, work by Parásará (from about the time of Christ) flowers are classified into four types on the basis of the position of the ovary and the form of the corolla. A contemporaneous work is by Agnivésa. The extent of botanical understanding in these texts is very remarkable, predating, western renaissance botany by 1500 years, and in many cases exceeding all western botanical understanding prior to the 18th century. There is a sophisticated understanding of plant anatomy, morphology and physiology. Even cell and tissue structure is understood, all the more remarkably because of the lack of a microscope. In addition Parásará recognized distinct plant families. These works though are only part of a long tradition of botanical study in earlier Vedic texts. Parts of this Vedic knowledge were carried with the spread of Buddhism throughout Asia.

Nestorian Christians driven from the Eastern Roman Empire had carried the Greco-Roman tradition into Persia. At Jundeshapur a university was established about 500 C.E. which kept the western tradition alive. The spread of Islam, and the trade and commercial prosperity that came with it, provided cultural links between Persian learning, and that in India and China, and also spread the oriental knowledge back to the west. Arab herbalists both preserved the classical herbalism of the west and added to it. The 'Canon of Medicine' of Avicenna (ibn Sina) (980-1037 C.E.), the foremost Islamic philosopher, included many plants unknown to Dioscorides. Avicenna came from the region of Iran. Meanwhile in the Islamic centers of Sicily and Toledo Jewish scholars translated the classical works. It was these tortuous roots, reaching back to the ancient knowledge of the Greeks and Indians, and elaborated by Islamic science, that fed the flowering of renaissance botany in northern Italy.

The Botanical Garden as a Model of Eden

Botanic Gardens played a pivotal role in the rise of a western commercial materialist hegemony and yet in their origins and early development it was their spiritual, religious and psychological meaning that was important. In Judeo-Christian-Islamic mythology the first garden was the garden of Eden. Plans of perhaps the earliest renaissance botanical garden at Padua show it as a circle representing the world with streams running out in four directions representing the four rivers that originated in the Garden of Eden. Other cultures and religious systems also had an image of an earthly paradise,. Gardens harked back and sought to recreate this paradise. The word for paradise comes from the Persian word *pairadaeza* for a walled park or garden but the word described something much older concept. Tomb paintings and carvings in Egypt, dating back to 5,000 years ago, show gardens; a formal pool containing lotus, fringed with papyrus, and shaded by palms and figs. Gardens were laid out in front of houses, around temples tombs. One painting records an expedition in 1,480 BC to the south the collect perfume providing shrubs for Queen Hatshepsut's garden at Thebes. Trees grown in gardens were fig, date palms, pomegranate, jejube, acacia, tamarisk, willows and nut trees. A broad range of other plants such as daisies, cornflowers, roses, irises, myrtle, jasmine, poppies, narcissus, *Persea*, jasmine and lotus were grown. Gardens provided shade and perfume. The arrangement is generally depicted as formal with rows of trees and pergolas supporting climbers.

The parks and gardens of Mesopotamia are not just legendary, they may have provided the inspiration for the description of the Garden of Eden. The Hanging Gardens of Babylon, constructed in the reign of Nebuchadnezzar II in the 6th century BC, were artificial, terraced and irrigated hills. The gardens were built for Nebuchadnezzar's wife Amyitis daughter of the King of Medea, perhaps to remind her of the hills of her home. The terraces were clothed in trees and plants recreating a sacred grove. They were probably irrigated by use of chain pumps. Several centuries later they were described by several Greek writers. Diodorus Siculus relates that they were 400 ft. in length and breadth and 80 feet high and built on foundations of stone. "The approach to the Garden sloped like a hillside and several parts of the structure rose from one another tier on tier. On all this, the earth had been piled and was thickly planted with trees of every kind that, by their great size and other charm , gave pleasure to the beholder. The water machines [raised] the water in great abundance from the river, although no-one outside could see it." The archaeologist Robert Koldeway claimed to have found the foundations of the gardens in 1899.

The gardening tradition lived on. Epicurus (341-270 BC) advocated the avoidance of pain and seeking of pleasure as the supreme goal in life and also made a famous garden just outside Athens beside the canalized Eridanus river. His students were known as "the philosophers of the garden" because he instructed them there and after his death they continued to meet there. The garden became a symbol of hedonism and detachment. As the 19th century poet George Herbert equated wisdom and nature.

"That Garden of sedate philosophy
Once flourished, fenced from passion and mishap,
A shining spot on a shaggy map;
Where mind and body, in fair junction free,
Luted their joyful concord; like the tree
From root to flowering twigs a flowing sap.
Clear Wisdom found in tended Nature's lap
Of gentlemen the happy nursery."
although his is anachronistic.

In the 16th century there was a renewed appreciation of the beauty of flowers. Plants were no longer recorded and collected primarily for their uses but out of intrinsic interest and for their beauty. Francis Bacon in the Elizabethan era had written that the creation of a beautiful garden was the highest form of artistic achievement. The publication in 1629 of John Parkinson's *Paradisi in Sole Paradisus Terrestris* was influential in encouraging the placing of plants in a garden landscape. Botanical gardens were being established across Europe to grow exotics and rarities but now there was a resurgence of an ancient idea, that the garden should be an image of the garden of Eden. Royal and aristocratic enthusiasts employed a succession of famous gardeners to create an earthly paradise for them. They introduced exotic plants from abroad and encouraged the cultivation of rarities from home.

The first modern botanic garden was established in Hamburg in 1316 but the widespread establishment of botanic gardens was inaugurated by the foundation of two at Padua in 1543 and at Pisa in 1545, the latter by Cosimo I de Medici on the urging of Luca Ghini. As well as founding the first botanic garden in Padua, Ghini was perhaps most influential as an enthusiastic teacher and correspondent with other botanists throughout Europe. One simple but lasting influence was his popularization of the collection and exchange of pressed dried plants, herbarium specimens. One correspondent was Aldrovandi in Bologna whose collection of pressed specimens is perhaps the earliest organized herbarium. Aldrovandi (1522-1607) was the first director of the botanic garden founded in Bologna in 1567. Others followed in Florence (1550), Leiden (1577), Montpellier (1593), Heidelberg (1597) and Copenhagen (1600). In the seventeenth century they were founded in Paris and Uppsala. The oldest surviving botanic garden in Britain, the Oxford Botanic Garden, was founded in 1621. The gardens provided a *raison d'etre* and the material of study. Many botanic gardens were initially as physic gardens. Padua was founded by the Senate of the Venetian Republic in order to train Paduan doctors on the urging of Francesco Bonafede, holder of the chair of *Materia medica* (*Lectura simplicium*), itself only founded in 1533. This is also the case in Tubingen which was founded by Fuchs in the 16th century and expanded as a proper physic garden in 1663. Chelsea Physic Garden in London was founded in 1673.

The explorers

The story of the study of flowers would be incomplete without the tales of the many botanical explorers and adventurers. There is no room here to tell even some of the stories of the rakes, buccaneers and libertines, as well as heroic eccentrics and missionaries, these botanists were. Only a few examples will have to suffice.

William Dampier (1652-1715) ran away to sea and ended up as a pirate. A flower in one hand and a sword in the other, between episodes of looting, rape and murder, he charted the Tropics and Antipodes, all the while collecting plants and making botanical observations. A despicable and sadistic drunk and he was once marooned by his crew in the Nicobar Islands, but paddled alone the 200 miles to Sumatra in a native canoe. On his return to England he was lionized and given command by the Royal Navy of an expedition to the Antipodes. His piratical past did not dismay the Admiralty, but what was remarkable, in that time of rum, sodomy and the lash, was that, on his return, he was court-martialed and convicted for cruelty, probably because it was to a subordinate officer and not an ordinary seaman. On the same voyage Alexander Selkirk, the prototype Robinson Crusoe, was marooned on Juan Fernandez Island. Years later Dampier, now employed only as a pilot of a privateer, was present when Selkirk was rescued.

The 18th century is notable for the state sponsored voyages of exploration. The achievements of James Cook with his accompanying botanists Joseph Banks and his assistant Daniel Solander, both the Linnaean apostles, are widely known. Banks was favored by an independent income, giving him the time and resources to support many botanical ventures. Having suffered many hardships on Cook's circumnavigation of the world and collecting many new species he settled in his house in Soho, which became a required call of any botanical notable. There was considerable Compton between states and individuals. One coup was the purchase of the Linnaean collection and library by James Edward Smith in 1788. It is now housed in a strong room, fitted out like an 18th century study, in the basement of the Linnaean Society of London off Piccadilly. One of Bank's schemes was the ill-fated introduction of Bread-fruit from the Pacific to the West Indies on the Bounty, captained by Bligh. Bank's botanist on the trip David Nelson died shortly after surviving the 4,500 mile journey in an open boat, and after seeing 1,000 of his carefully nurtured breadfruit seedlings dumped overboard by Fletcher Christian's mutineers.

Another of Bank's protégés was more successful. Francis Masson introduced many heathers, pelargoniums and other plants from South Africa but after surviving capture by French privateers and hostile natives, Boers and escaped convicts, he too died young, frozen to death in North America at the age of 33.

The story of Philibert Commerson, on the rival French expedition to Cook's by Bougainville, appeals because it includes more of the kind of lunacy and obsession that accompanies many of the adventures of these botanical maniacs. Philibert Commerson (1727-1773) was trained at Montpellier where he got into trouble for stealing plants for his herbarium from the gardens of residents and even the botanical garden. He made his name as a heroic plant collector, escaping one mountain avalanche by rolling down ahead of it like a ball. Falling into a mountain ravine, on another occasion, he was trapped by his hair in a bush. Cutting himself free meant that he fell into a raging torrent and nearly drowned. Commerson seems to have taken up the chance to travel with Bougainville's expedition with alacrity, as if he was desperate to get away from home. Accompanied by a faithful assistant, a young fresh-faced lad called Jean Baret, he travelled halfway around the world collecting specimens and making notes. Things went awry in the New Hebrides when a chieftain took a fancy to Jean Baret. He felt him up and disturbed his clothes, and Jean was revealed to be a Jeanne, actually Commerson's housekeeper from home. She had chased

Commerson to his ship and there taken disguise in order to be with him. It is too much to suggest that Commerson refused to go back to France in embarrassment, because he was signed off by Bougainville in Mauritius and joined Pierre Poivre and made a profound contribution to the development of a new botanical consciousness.

Another botanical nutcase was David Douglas credited with introducing 200 species to Europe from North America including the giant conifers such as the one that bears his name the Douglas fir (*Pseodotsuga menziesii* – the specific epithet recording the efforts of Archibald Menzies before him). In the third decade of the 19th century Douglas survived a bolting horse that only understood commands in French, having all his belongings stolen while he was climbing a tree to collect seeds, encounters with bears, falling down ravines, cold and starvation as well as several episodes of capsize, only to die trample to death in a pit constructed to capture feral cattle on Hawaii.

Not all collectors suffered such hardships all the time. Joseph Dalton Hooker collecting in Sikkim in 1848 was accompanied at first by a retinue of 56 porters and servants. By that time botany had become an imperial state-sponsored activity.

The renaissance of botany

A renewed interest in classical texts was part of the renaissance. Theophrastus' work disappeared from view in the Middle Ages of Europe though it remained an influential text in the Islamic world. Pope Nicholas V instigated the translation by Theodore Gaza (1400-75) of copies discovered in the Vatican library. A new translation of Dioscorides was made by Pietro d'Abano in 1478. At the same time as Theophrastus and Dioscorides were becoming more available there was also a realization of the limitation of the Classical authors. Scores of new plants were being brought back to newly founded botanic gardens. Artistic developments encouraged the more realistic depiction of nature seen to perfection in masterly drawings by Da Vinci and Durer. Italian manuscript herbals of the 15th century showed a renewed appreciation of the value of accurate drawings of plants. The introduction of printing with crude woodblock diagrams, usually derived from debased manuscript copies, came as a temporary setback, but by the early 16th century detailed and beautiful woodcuts were being produced.

Outstanding among these are the woodcuts made from the water-colors of a pupil of Durer called Weiditz. They were used to illustrate a Herbal by Brunfels called the *"Herbarum Vivae Eicones"* the first volume of which was published in 1530. The text was little more than a pedestrian plagiarization of previous authors with a few new snippets but Weiditz's woodcuts are marvelous in their vigor and accuracy, though some look as if the flower has been left out to wither on the artist's table for too long. Brunfels was assisted by Bock who produced his own text the *"Kreütter Büch"* in 1539. At first it was produced without figures because of the expense but this may have been fortuitous because their lack may have encouraged Bock to describe the plants in detail, including their parts, their life history, and in a clear and colloquial German. Another outstanding work was that of Fuchs. His *"De Historia Stirpium"* published in 1542. The text owes something to Brunfels and Bock , though it is clear that Fuchs was a good field botanist. However, like Brunfels work it is the figures, by Meyer and Füllmaurer, which capture the attention. The importance of Fuchs and Brunfels was recognized later by their names being given to the genera *Fuchsia* and *Brunfelsia*. though perhaps these genera should have been called *Meyereana* and *Weiditzia*.

Gardens were stocked by an ever-widening diversity of flowers collected from around the world. The herbarium and the botanic garden became the two pillars of systematic botanical research and are still essential today. It is hard to imagine what kind of botanical scientific progress would have been possible without them. An outstanding collector was Ghislein de Busbecq the Hapsburg ambassador to the Court of Suleiman the Magnificent in Constantinople between 1555-62 . The previous ambassador had been imprisoned in the Sulatan's dungeon and had returned to Vienna a broken man. Three of the plants Busbecq introduced to northern Europe were the lilac (*Syringa*), horse chestnut (*Aesculus*) and tulip (*Tulipa*). Each entered our cultural landscape in a remarkable way.

The religious troubles of the 16th century touched the lives of a series of botanists. P. Mattioli was as good a self-publicist as a botanist. As a young man he was a protégé of Ghini, a debt he admitted to in a letter to Aldrovandi after Ghini's death, and he was lucky to have as his friend Busbecq who supplied him with new plants and information from the east. He did produce the first vernacular translation of Dioscorides and gained fame as a result of the publication of his verbose commentaries, *"Commentarii in VI Dioscorides Libros",* in 1554 which sold 30,000 copies, perhaps largely because of the excellent illustrations by Leberale. A translation into Latin with the synonyms in several languages for plant names was a great advance popularized Dioscorides. 50 different editions in many different languages followed. Self-important and spiteful, Mattioli managed to get a rival called Luigi Anguillara sacked from his Professorship at Padua. Another critic, Amatus Lusitanus, Mattioli called by Mattioli "Amathus" from the Greek for stupid, suspect as Christian convert from Judaism, was reported to the Inquisition - and had to take refuge in a Turkish ghetto, losing all his possessions in the process. Another of Mattioli's unpleasant side-lines was to test the toxicity of *Aconitum napellus*, monkshood, by feeding it to condemned criminals.

The achievements of Mattioli's contemporary William Turner, sometimes styled the Father of English Botany were more modest, though the man was just as remarkable. Of humble origin, the son of a tanner, he published the first original botanical works on the English Flora including *"The Names of Herbes"* (1548) and *"A New Herball"* (1551-68). The illustrations came from Fuchs. Turner was close friends of Ridley and Latimer at Cambridge, who were later to be Protestant martyrs. An outspoken Geordie, he was a scourge of bishops and fancy priests, even training his dog to snatch and run off with their mitres. Forced to leave the country in Henry VIII's reign he travelled widely on the continent. Returning to become the Dean of Wells on the accession of Edward VI, he had to escape abroad again on the accession of Mary I, this time taking the opportunity to attend Ghini's lectures in Italy.

During his continental excursion Turner met and became a lifelong friend of Gesner (1516-1565), an outstanding botanical collector, and a remarkable botanical illustrator, though he never saw his work published. He produced 1,500 drawings, which far surpassed the achievements of others, because they were drawn from living material, later on from the botanic garden in Zurich that he established in 1560. He included separate detailed diagrams of flowers, fruits and seeds. They were converted to woodcuts at his own expense. A few were utilized by other authors, but without acknowledgement, but the bulk were not published until 1751. Nevertheless Gesner was had great influence on the development of botany through his contacts with other botanists. Part of Gesner correspondence was published by Jean Bauhin in 1591 with whom he botanized with in Switzerland. Gaspard Bauhin his brother, nearly 20 years younger published his *"Pinax"* in 1623. It was innovative in including comprehensive references to previous works and a list of names thought to be synonymous. Probably influenced by Gessner, he recognized genera each of which could be split into a number of species G. Bauhin frequently used a binomial name for each species, using a generic part and a specific descriptor.

In Paris, John Robin (1550-1629), and his nephew Vespasian (1579-1662), grew plants from North America. Their Paris garden was to later transmogrify, much enlarged into the Jardin des Plantes. The black locust introduced from North America, but probably first grown in the Tradescants' garden, was given the name *Robinia pseudoacacia* in their honor. They exchanged plants with John Gerard who had a notable garden in Holborn in London.

There was a triumvirate of notable Flemish botanists and friends whose lives were variously disrupted by the religious conflicts at the end of the 16th century. Rembert Dodoens (Dodonaeus) lost his home and possessions in the Spanish wars but eventually became the personal physician to the Holy Roman Emperor, a post Mattioli had once held. Charles de l'Ecluse (Clusius) lost an uncle in one Spanish massacre and his father had to flee to his son's protection in Antwerp. Nevertheless Clusius traveled widely, studying under Luther and Rabelais, collecting in Turkey and Spain new varieties of daffodil, and fritillary among many others species. He was instrumental in the introduction to northern Europe of the potato from Spain Impoverished he had a change of luck in 1573 when he became supervisor of the Imperial Gardens in Vienna and then in 1588 professor at Leyden. Mathias De l'Obel (Lobelius) lived variously in France London and Antwerp. For a while he was the personal physician of William the Silent and later became the botanist of King James I in London. Dodonaeus, Lobelius and Clusius are now remembered because of the genera they gave their names to; *Dodonea* (ebony), *Lobelia* and *Clusia* but they worked extensively with the Antwerp publisher Plantin whose main artist was called Borcht, another refugee from the religious wars.

John Gerard translated Dodonaeus' work into English and published it in 1597, with a few additions, as his own *Herbal*. The translation, which may not even be Gerard's work but that of a man called Priest who died before he had finished, was inaccurate, though stylish, and even the illustrations were lifted - from a German Flora. Lobelius was called in by the publisher to make thousands of corrections, but quarreled with Gerard, who accused him of not understanding English. It could not help that Lobelius found that he too had been plagiarized. Gerard did describe some new plants, but even his publisher noted that Gerard's new discoveries of native plants were a bit suspect. The peony he accidentally found growing wild near Gravesend in Kent he had planted there from his own garden.

As botanical gardens were being established across Europe to grow exotics and rarities, royal and aristocratic enthusiasts employed a succession of famous gardeners to create an earthly paradise for them. Parkinson was given the title *Botanicus Regius Primarius*, the King's Principle Botanist by Charles 1. A neighbor of his in Lambeth was Tradescant, gardener I succession to Robert Cecil and George Villiers Aristocratic collectors competed by financing collectors to improve their stock. John Tradescant laid out a garden for Robert Cecil Earl of Salisbury at Hatfield house and later served the first duke of Buckingham and Charles I. His own garden in Lambeth became a treasury of exotic plants from abroad. Tradescant made repeated trips to the continent to collect plants and also accompanied a diplomatic mission to Muscovy and an expedition to the Barbary coast. The opening up of the New World provided an immense source of new plants - many of which could grow very well in the gardens of north-western Europe. Tradescant's son also called John traveled to Virginia three times and to Barbados. Together they introduced via their Lambeth nursery exotic plants like the Virginia creeper (*Parthenocissus quinquefolia*), Tulip tree (*Liriodendron tulipifera*), Michaelmas daisies (*Aster novibelgii* and others), lupins and the spiderwort, that was named for them, *Tradescantia virginiana*.

Meanwhile a field botany tradition was being established. John

Ray (1627-1705) gained his habit of long botanizing country walks after a boyhood illness. Ray became an established scholar at Trinity College Cambridge, but the impetus for more extensive travels came about because of the Act of Uniformity of 1662 that required all college fellows to take an oath accepting the Common Book of Prayer. Ray refused to sign and took to the road. Luckily he had a wealthy friend, also a member of college, called Willoughby. Together they toured the British Isles and the continent.

The close connection with North America continued throughout the 17th and 18th centuries. Philip Miller at Chelsea Physic Garden in London was eager to receive seeds from the Quaker farmer John Bartram. John Bartram (1699–1777) began his botanical exploration and collection at about the time of his first wife's death in 1727 and in 1732 began a correspondence with a wealthy fellow Quaker, London cloth merchant and plant enthusiast . He established an ornamental garden and each year after the harvest made an expedition to collect new plants. He exchanged plants and seed with Collinson, who in time became his patron and agent, selling on material to a range of collectors, including Linnaeus, Sir Hans Sloane, a bevy of aristocratic enthusiasts and King George III, for whom in 1765 Bartram became the "King's Botanist" on a stipend of £50 per year. The stipend financed an excursion to Florida that had recently become a British dominion. Linnaeus is said to have called him "the greatest natural botanist in the world." Between 150 and 200 North American plants were introduced to Europe through Bartram. His sons John and William Bartram carried on the tradition with William making expeditions to the west.

The collecting maniacs

Botany was no insipid pastime but a vigorous and exciting adventure. Rumbustious botanical excursions from the Chelsea Physic Garden caused a public scandal. Abroad botanical collectors were among the first exploring new regions. At home it was the avaricious competition between collector's that stoked a collecting mania.

Ogier Ghislain de Busbecq brought tulip bulbs back to the Imperial garden in Vienna in 1554. The word tulip is a corruption of a Turkish word for turban-shaped. The Turks had selected from the wild especially pretty variants. Varieties with long pointed petals were particularly favored. However, it was Clusius experimenting in the Imperial Garden in the years 1573 to 1589, with crosses between varieties to create new cultivars that first stimulated the tulip fad. Clusius moved to Leiden in 1593 to establish a new physic garden, bringing some bulbs with him, and intending to continue his crossing experiments, but by accident provoking the avariciousness in his colleagues. Many of his bulbs were stolen and soon they had been widely propagated. They had lost their value for Clusius and he gave up their cultivation, but about 1608 a craze started in France, that spread north to Holland. By 1629 Parkinson, in perhaps the first great garden book, the *Paradisi in Sole Paradisus Terrestris*, numbered 140 different varieties of tulips in English gardens.

Between 1634 and 1637 there was a frenzy of speculation in tulip bulbs. Huge sums were risked on the newest varieties. Many streaky varieties were the result of viral infection. One bulb cost a carriage and pair, another twelve acres of land. Semper Augustus, a red tulip with white streaks sold for 10,000 guilders, enough to buy a grand canal-side house. In Haarlem one merchant bullied a rival into selling a duplicate bulb for 1,500 guilders and then stamped on it to preserve the value of his one. A seaman accidentally mistook another bulb worth 3,000 guilders for an onion and ate it. He ended up in jail for 6 months. Soon 'paper' tulips, futures on potential tulips, were being traded, but in 1637 the bubble burst ruining many speculators. However the enthusiasm for flowers did not disappear.

The Turks themselves had their own tulip craze in the reign of Sultan Ahmed III between 1703-1730, known as the tulip era. The Sultan imported bulbs from Holland and also Iran to add to his collection of native tulips. Huge prices were paid. During the flowering season the imperial gardens held a tulip festival. The garden lit by candles, some carried around on the backs of giant tortoises, eunuchs bearing torches, mistresses from the harem dressed to show off the blooms, and song birds in gilded cages all added to the magical effect.

In the 19th century Veitch grew to be the biggest family firm in Europe and by the First World war had nearly1300 cultivars for sale providing for a succession of fashions and crazes and commercial nurseries were established to supply them. The availability of plate glass allowed the development of hothouses, greenhouses and conservatories. Now the gardener was not limited to plants that could be grown outside in temperate Europe and North America. Orchids were a particular fad.

Dahlias from Mexico were first introduced by Menonville to Europe accidentally, as food for smuggled cochineal bugs. The bugs didn't survive but André Thouin grew up the plants at the Jardin de Roi, hoping the tubers would provide an alternative to potatoes. Unfortunately the experiment was abandoned because of their bitter taste and successful introduction had to await the raising of plants from seed in the Royal Gardens in Madrid by Abbé Cavanilles, who named the plants after Dahl, a disciple of Linnaeus. Dahlias were first grown successfully in Britain in 1804 and hundreds of varieties were developed in the craze that followed, and £1000 was offered for a blue dahlia.

The Victorian fern craze flourished from the 1840s onwards. As railways made the countryside more accessible people of all classes made trips to collect them, and ferns from diverse parts of the British Empire were set home. In 1851 Charles Kingsley coined the term "*Pteridomania*" to describe it. Another craze was for rhododendrons. Orchid maniacs maintain the manic collecting tradition.

Systematizing nature

Cesalpino was the first (1519-1603) great plant systematist. He was a student of Ghini's and also corresponded with Gessner. Through his work these two great botanists gained a lasting influence. Cesalpino, taught at the university of Pisa for nearly 40 years. He did not endear himself to the Inquisition with his original views but eventually he became the physician to the Pope. He took up Ghini's focus on the recognition of species within genera and made many other advances in botany, human anatomy and medicine. "Genus" from the Greek *genos* was a concept of Aristotle who also described forms "*eidos*" of the genus, the progenitor of species, but these concepts had not been precisely applied. *De Plantis*, published in 1583, provided the first new comprehensive classification of plants since Theophrastus. Remarkably Cesalpino attempted a natural classification and rejected classification by the uses of plants or in an alphabetical arrangement favored by the herbalists. He made a clear distinction between essential or fundamental characters and accidental or superficial characters which are not stable and are likely to change depending upon the climate or soil.

John Ray (1623-1705) also developed a system of natural classification which grouped together plants on the basis of their natural affinity. Like Cesalpino he rejected accidental characters. His works *Methodus Nova* (1682) *Historia Plantarum* (1686) and *Methodus Emendata* (1703) were remarkable for their sophisticated analysis of variation plants. He coined or popularized several terms, including petal, cotyledon and pollen. Ray was an Essex blacksmith's son who managed to get a Cambridge fellowship. He botanized widely in Britain, throughout the Civil War period, sometimes alone on horseback, and sometimes in the company of his young aristocratic friend and sponsor Willoughby. Like Turner before him, his uncompromising Puritanism meant an enforced tour of the continent when he lost his Cambridge fellowship on the Restoration of Charles II in 1662. As well as writing the first proper Flora of the British Isles it was Ray's intensely practical approach which allowed him to reach new conclusions from his own observations. Especially influential was the establishment of the major groups of plants like the monocots and dicots - plants with either one or two seedling leaves and the enangiosperms (angiosperms) and gymnosperms - plants with and without enclosed seeds, though he was confused about plants with 1-seeded fruits which he put in the gymnosperms. These distinctions had already been established by Parásará in India 1500 years before but were unknown in the west.

Pierre Magnol (1638-1715) Professor of Medicine and later director of Montpellier botanic garden met Ray in 1664 and was greatly influenced by him. Using Ray's method Magnol defined a series of families, including the Ranunculaceae, Papaveraceae, Papilionaceae and Malvaceae which are recognized today. Tournefort's (1656-1708) work was in some ways less advanced than Ray's but he made a valuable contribution to botany by providing brief descriptions of genera so that genus became firmly established as a rank in the taxonomic hierarchy. He followed in Busbecq and Clusius' footsteps and brought back 1,300 specimens from the Levant but he was also inclined to raid the gardens of friends and strangers for specimens. Through his *"Élémens de Botanique"* (1694) and the *"Institutiones Rei Herbariae"* (1700) he became perhaps the most widely read member of the Parisian Academy of Sciences outside France. In 1683 became the professor of botany at the Jardin Royal in Paris. One of his outstanding pupils was Vaillant (1669-1722) who built the first greenhouse in France.

Gradually a taxonomic hierarchy a series of less and less inclusive categories was becoming established. It proved to be an effective way of representing the pattern of relationships between species. For example some of the important categories for a species of columbine are

Class Magnoliopsida
Order Ranunculales
Family Ranunculaceae
Genus *Aquilegia*
Species *A. vulgaris* L.

On average two species in the same family are less closely related than two species in the same genus.

At about the same time the binomial system of naming species became established whereby a species name consists of its generic name plus a specific epithet. . The trouble with previous methods of naming species was that they were really a means of identifying the species. Some names consisted of a whole sentence of descriptive terms and were very cumbersome. Bauhin had used binomials but these were constantly being superseded as new species were discovered and new characters had to be used to distinguish the known species. Linnaeus had the lucky thought that with a workable classification for identification the second name of a binomial could act merely as a trivial label fixed for all time to that species; it did not describe the species , it merely labeled it. This introduced a stability into naming species which was vital for scientific communication. The species called *Geranium columbinum majus dissectis foliis* by Ray became *Geranium molle* L.

But how was the botanist to find the name? The sexual system of classification of Linnaeus answered that problem too by providing a ready mechanism by which all known plants could be identified. It was a system which relied only on the ability to observe the male and female parts and to observe some very simple features of them like their number. There were 24 classes based on the number of stamens, their relative length and their degree of fusion. Each class was subdivided into orders which differed in the number of pistils.

The empire of botany

"The greatest service which can be rendered any country is to add a useful plant to its culture." Thomas Jefferson

From the earliest times there has been a drive to obtain dominion over plants, to collect them and cultivate gardens, to understand and categorize them, and not just because of the materials they provide. The variety of plants has been a constant enthusiasm for humans and the study of plants predates civilization. Gardeners and botanists have sought to possess and understand plants. Throughout history they have obsessively sought out new varieties, to collect them, either to own them jealously, or to share them. Scientific botanists have sought to take dominion over them by understanding them, cataloguing them and dissecting them. The knowledge and possession of plants was a power, a power used by the agronomists and imperial botanists, to introduce crops in new areas, to increase productivity and to take dominion of over the land.

This is seen most clearly in the role of botanic gardens in the imperial botany of western colonial expansion. It was the European empires who changed the world by introducing plants to new areas, the first step in what has come to be called globalization, the homogenization of the world and its economic resources. Tropical crops were introduced into new tropical regions and European crops and their weeds were widely introduced everywhere. The Portuguese were the earliest European explorers to reach the east by sea. first, and the other powers soon followed. As trading posts became more established as colonies to exploit and develop natural resources the movement of plants became more systematic.

The Dutch established botanic garden at the Cape and the French one on Mauritius. A botanic garden was founded in St Vincent in 1765. Calcutta botanic garden in 1787 by Colonel Robert Kyd. In seeking financial support for the garden Kyd appealed to the directors of the East India Company,

"I take this opportunity of suggesting to the Board the Propriety of

establishing a botanical Garden, not for the Purpose of collecting rare Plants (altho' they also have their use) as things of mere curiosity or furnishing articles for the Gratification of Luxury, but for establishing a stock for the disseminating such articles [as] may prove beneficial to the Inhabitants [of India], as well as Natives of Great Britain and ultimately may tend to the Extension of the National Commerce and Riches."

At home in the UK John Lindley chaired a Treasury Commission that urged the foundation of a National Botanic Garden to co-ordinate the activities of the botanic gardens scattered through the British colonies.

"A national garden ought to be the center round which all minor establishments of the same nature should be arranged ...receiving their supplies and aiding the Mother Country in everything that is useful in the vegetable kingdom. Medicine, commerce, agriculture, horticulture, and many valuable branches of manufacture would benefit from the adoption of such a system. From a garden of this kind, Government would be able to obtain authentic and official information on points connected with the founding of new colonies; it would afford the plants required."

With this emphasis on economic botany Kew Gardens were transferred to the state in 1841 with the first of the Hooker dynasty, William Jackson Hooker, as its first director . The transfer of plants was

The unauthorized piracy of rubber seeds was not the last of Kew's activities, as an imperial agent. Indeed the appointment of Thistleton-Dyer as Director in 1885 brought these activities to the fore. He had supervised the propagation of rubber plants in the Peradeniya Gardens in Ceylon (Sri Lanka) in 1876 and cacao 1880. As director he regularized the links between Kew and the colonial gardens and closely co-operated with the Colonial and Indian Offices . In his introduction to Bean's history of Kew published in 1908, after he had retired and had been replaced by Prain, former director of Calcutta botanic garden, Thistleton-Dyer wrote

"There are some sixty district governments under the British Crown

and in any technical difficulty all have a resort to Kew. It did what was possible when coffee-leaf disease brought financial disaster to Ceylon; the fortunate identification of a single leaf started the rubber industry of the Gold Coast; Kew sent tea to South Africa; it gave cinchona to India, and a dose of quinine can be purchased at any Indian post office; it transferred the South American rubber plants to the East, with results which have been described as fraught with 'wealth beyond dreams of avarice.' A chain of Kew-trained men dot the course of the future Cape-to-Cairo railway. Scientific members of the Kew staff hold important positions in India and the Transvaal, and a former assistant director has done noble work in restoring agricultural prosperity in the West Indies."

As Brockway has pointed out perhaps the most important thing Thistleton-Dyer did was to establish the Kew Bulletin of Miscellaneous Information. This heralded the development of a new kind of imperialism, that of western scientific knowledge. Through the pages of this journal, commercially valuable information was made available to western governments at the expense of established plantation industries.

Imperial botany brought immeasurable benefits but it also exploited terribly native peoples. The most notorious example of imperial botany was the exploitation of wild rubber instigated by King Leopold II of the Belgians at the end of the 19th century . In what he called Congo Free State, now the Democratic Republic of Congo, he carved out his own personal colony. He employed H. R. Stanley, the journalist who had found Livingstone and who between 1874 and 1877 had traveled the depths of the Congo. Stanley bribed 450 chiefs to sign over their lands and their people to Leopold. At the Berlin conference held in 1884-5 the major powers granted the Congo to Belgium; it was a useful way of avoiding big power rivalry over the territory. The native people were required to trade only with Leopold's agents. When by 1890 it was clear that they refused to cooperate the agents were paying chiefs to provide labor, or buying or stealing slaves from Arab slave traders to establish a *Force Publique*, a private army of 16,000, with 350 European officers, to extort the native tribesmen.

A boom in rubber prices following the invention of the inflatable tire encouraged further atrocities. A tax was imposed of chieftains to require them to provide labor. Slave traders were eliminated but instead the Force Publique pushed deep into areas where the "rebels" had retreated to avoid the tax, burning villages and slaughtering villagers, including women and children. Soldiers were ordered collect the right hands of "rebels" that they had killed to prove that they had been doing their job, but it was easier to cut off the hands of the living, and it had the additional advantage of terrorizing the population into submission.

It was the campaign of Edmund De Morel a British shipping clerk working in Antwerp who first alerted the British government to what was going on. He had noticed how the trade was one-way and could only have been sustained by systematic extortion. They sent Roger Casement, later to be a hero of the Irish rebellion, but then a British diplomat, to the Congo to investigate, where he uncovered a history of murder, hostage-taking, mutilation and forced labor. He estimated in his report that 3 million Congolese had died as a result of the system, from shooting, starvation or disease that were the result of the terror system. Later historians have much higher estimates of around 8 million deaths! Together with De Morel, Casement founded in 1904 the Congo Reform Association which campaigned relentlessly in Britain and the USA. By 1908 Leopold had turned over his Congo fiefdom to the Belgian government leading to a cessation of the worst of the excesses, he died in 1909, and anyway soon attention was directed elsewhere.

First the exploitation of Cinchona, then the rubber boom was leading to the exploitation of the native people of northwest Amazonia, in the region of Putumayo. Here racial and tribal divisions were exploited to provide the labor to gather latex from wild caucho or uli (*Castilla elastica* and *C. ulei*) and jebe (*Hevea*) species and process it into balata balls . For example The Peruvian Amazon Company, a British company founded by Casar Arana, enslaved and brutally treated its workers. Casement claimed in 1912 that 30,000 Indians had been killed and thousands enslaved and tortured. For a while the wild rubber trade dominated world trade but ironically in a few short years rubber from Asian plantations became available in large quantity destroying the Amazonian and Congolese trade.

Say it with flowers

While governments exploited the world's plants for commerce, for the people plants kept a deeper psychological and symbolic meaning. "*Say it with flowers*" was a slogan coined for the American Society of Florists, but the *Language of Flowers* was quite literally an attempt to systematize the symbolic character of flowers. Charlotte de Latour in a book published in Paris in 1819 codified the language of flowers by providing a dictionary of meanings. It is all very amorous and very ridiculously detailed. A rosebud with its thorns and leaves signified "I fear, but I am in hope" while without its thorns it meant "there is everything to hope for". Without its leaves meant "there is everything to fear" when upright but upside down meant "One mustn't fear or hope". A marigold on the head meant trouble to the mind, on the heart, the pain of love and on the breast boredom. The whole nonsense, which became extremely popular, and was elaborated by many other authors was given the gloss of being an ancient science practiced in classical times and the Orient.

A different symbolic meaning has been given to the same flowers over the ages. In Shakespeare's Hamlet the deranged Ophelia says:

"There is rosemary, that's for remembrance.....
and there is pansies, that's for thoughts....."

and then she gives fennel to the King signifying flattery and columbines to the Queen signifying adultery. The symbolism is poetic and complex. Three flowers are supreme in their symbolic weight: the lotus, the rose and the lily.

Lotus buds, rosettes and palmettes are a recurrent decorative theme both in Egypt and in Mesopotamia. The lotus, *Nymphaea coerulea* is the blue lotus which was sacred in Ancient Egypt. Its flowers open at dawn and close at night and so it became associated with Ra the Sun god. The bud clearly had a phallic significance too. The flowers are fragrant. There is also a white lotus N. lotus which opens at night and also grows in the Nile delta. The Indian Lotus *Nelumbo nucifera* was introduced in the Persian period. In Egypt a heraldic lily, representing the south, was produced prolifically on sculptures from Old Kingdom times. The Madonna lily *Lilium candidum* is figured on Cretan frescos from 5 000 years ago. It is native to the eastern Mediterranean region but has long been cultivated for its white lowers used for making scent. White lilies are mentioned in the Bible. but some have suggested that the Rose of Sharon may be *L. candidum*. Another suggestion that this is *Pancratium maritimum* while the Lily of the Field is *Narcissus tazetta* or *Hyacinthus orientalis*. The sweet smelling lily was identified with springtime and rebirth in the Roman period. In the early Christian period it was rejected for being a symbol of luxury and idolatry but later it came to symbolize purity and was associated with chastity and the Virgin Mary.

Papyrus represented the North. Later *Acanthus* leaves provided the inspiration for the carved capitals of columns in Greece. Flowers provided inspiration for styles of personal adornment, like the beautiful necklaces, of blue and white lotus, daisies and cornflowers from Egypt, and the Sumerian caps, crowned with gold daisies, that can be seen in the British Museum.

The development of botany in China ran in parallel to that in the west. For over a millennium in the Middle Ages Chinese botany far surpassed western botany. There was a strong herbal tradition encouraged by bureaucrats and emperors. Over the centuries many different kinds of plants were included in Herbals. The earliest monograph on chrysanthemums dates from the beginning of the 12th century C.E.it lists 25 different cultivars. In contrast to the west, the tradition was for a high degree of accuracy in the description and figuring of plants. There was also a cult of flowers and gardening, even a peony mania in the early ninth century. Peach, chrysanthemum, lotus, peony, Cymbidium orchid and tiger-lily were all cultivated for their beauty from 1000 B.C.E. From about 600 C.E. many others like camellia and magnolia were also cultivated. Gardens provided a quiet retreat from the world where the beauty of nature could be contemplated. The arts of poetry and painting were intimately connected with the study of flowers. For example a poem describing the peony runs:

The deep green foliage is quiet and reposeful
The petals are clad in various shades of red;
The pistil drops with melancholy
Wondering if spring knows her intimate thoughts

The four noble plants were the plum, bamboo, orchid and chrysanthemum. Chrysanthemum, the flower of autumn, was the symbol of joviality and contentment. Bamboo for spring was the symbol of the courteous gentleman, and of companionship and modesty. The orchid for summer was the symbol of the refinement of beautiful women and great men. The plum, symbol of winter, was for chastity and feminine beauty and also had a complex number symbolism: the blossoms are Yang, heaven, the branches, Yin, earth. The pedicel is the Ridgepole of the Universe, the three sepals are Heaven, Earth and Man, the petals are the five elements and the stamens are the seven planets.

Floral imagery dates back to the Indus Valley civilization of 5,000 years ago. Hinduism had a strong identification with nature and the sacredness of plants. The lotus blossom had a paramount significance but many other plants like the bright yellow and orange flower garlands of marigolds were favored in ceremonials and also in private life. Garlands were a bed-time adornment of men as well as women. Jasmine flowers were used to dress the hair. Much of the imagery was sexual and poetic but running in parallel was a scientific study of plants.

Plants have had an important symbolic significance. Olive foliage was the sign of goodwill (the olive branch). Cherry laurel (*Prunus laurocerasus*) was used to crown the Caesars, perhaps quite appropriate considering their later murderous behavior, since the leaves are contain cyanide which is released when they are damaged. The rose, sweet scented and full of soft petals, was associated with luxury, with Venus, with love and spring, but also with death. Garlands of roses were placed as wreaths at tombs. This was justified in Early Christian times because with its thorns it also came to symbolize Christ's agony. Flowers as garlands or chaplets were incorporated into many religious ceremonies. The cult of the Virgin Mary resulted in many flowers being baptized with Christian names. Calendula (Marigold), *Alchemilla* (Our Lady's mantle), *Spiranthes* (Our Lady's tresses), *Cardamine* (Our Lady's smock), *Anthyllis* (Our Lady's fingers) are just some of them. Flowers also entered heraldic imagery. Broom (*Genista*) was the symbol of the Plantagenet dynasty (planta-genista). Some writers identify the Fleur de Lis (flower of Louis) as a lily but it is probably an *Iris florentina* the source of the violet-scented insecticidal and medicinal orris root.

Garden History

Towards the end of the 17th century and in the 18th century there was a change of sensibility. The clipped geometrically shaped symmetrical gardens of the Restoration with their symmetric parterres gave way to rococo asymmetry, which at its margins blurred the distinction between garden and nature, and then there was the exuberance, the wildness of baroque. Nature and the countryside was colonized again, but now by the intellect. Just as towns and cities were starting to burgeon like great excrescences, gardens were now to be constructed to recall an Arcadian landscape when life was pure and simple. They were to be artfully constructed to present a picture of an idyllic landscape.

The landscape artists, either the idealized landscapes of Lorraine or the romanticism of Friedrich and Constable, and in Japan the woodcuts of Hiroshige and Hokusai, have all trained our eye. The clipped geometrically shaped symmetrical gardens of the Restoration with their symmetric parterres gave way to rococo asymmetry, which at its margins blurred the distinction between garden and nature, and then there was the exuberance, the wildness of baroque. Nature and the countryside was colonized again, but now by the intellect. Just as towns and cities were starting to burgeon like great excrescences, gardens were now to be constructed to recall an Arcadian landscape when life was pure and simple. They were to be artfully constructed to present a picture of an idyllic landscape. The landscaped parks of William Kent, Charles Bridgeman, Lancelot "Capability" Brown and Humphrey Repton marked the final phase of taking possession of the landscape. Now the even the wild or semi-natural was to become reformed in an ideal image as part of the estate of the landowner. Landscaping was as much part of the idea of improvement as the agricultural developments of the 18th century. Jane Austin's applies her caustic wit against the improvers in Mansfield Park. "Improvement" marked just the latest stage in the appropriation of the countryside by a privileged few.

Improvement could be theatrical and painterly as at Stourhead, involving large scale works like the diverting of rivers and streams, the damming of lakes and the moving of mature trees. Gardening on the grand scale can be seen at Blenheim and many other grand country house. It might also include more subtle changes of landscape. Views were very important. Through the views the park seemed to include the countryside around. This could include, not just scenes of wilderness, with natural features incorporated or enhanced, but the productive countryside of fields and woods. Ha-has, hidden boundary banks and ditches were constructed so that the riff-raff were excluded but the lordship of the eye was unimpeded. In part agricultural development funded the development of landscaped parks but in its turn the timber of the landscape parks was an important source of income. Humphry Repton railed against the nouveau riche who only had a commercial interest but in time the improvement came to be seen more pragmatically as necessary investment for commercial return. There had always been some element of pragmatism. In 1713 Joseph Addison wrote in The Spectator

"Why may not a whole Estate be thrown into a kind of Garden by frequent Plantations, that may turn as much to profit as to the Pleasure of the Owner? A Marsh overgrown with Willows, or a Mountain shaded with Oaks, are not only more beautiful, but more beneficial, than when they lie bare and unadorned. Fields of Corn make a pleasant prospect, and if the walks a little taken care of that lie between them, if the natural Embroidery of the Meadows were helped and improved by some small Additions of Art, and the several Rows of Hedges set off by Trees and Flowers, that the Soil was capable of receiving, a man might make a pretty Landskip of his own Possessions."

Grand landscape design was in decline by the beginning of Victoria's reign, but by this time the Romantic Movement had taken psychological possession of the landscape and its flora. The countryside and nature became now as much a mental landscape as a real one. At its most shallow, this was expressed as a search, not just for beauty, but for the "picturesque", a scene capable of being painted. This was a way in which beauty could be circumscribed, described, possessed. It has provided the vocabulary of our appreciation of the countryside and nature. Jane Austin poked fun at the cult of the "picturesque" in her novel Sense and Sensibility published in 1797. Edward describing a walk to Marianne says

"You must not inquire too far, Marianne - remember, I have no knowledge in the picturesque, and I shall offend you by my ignorance and want of taste, if we come to particulars. I shall call hills steep which ought to be bold; surfaces strange and uncouth, which ought to be irregular and rugged; distant objects out of sight, which ought only to be indistinct through the soft medium of a hazy atmosphere."

Nature spake to me rememberable things

The Romantic Movement was a revolt in the late eighteenth and early nineteenth centuries against, among other things, the mechanistic Newtonian world view which was increasingly regarded as dry and sterile. At its root was the perceived loss of the spiritual dimension of nature and humans. It was characterized in literature, music and painting by freedom of form, and creative imagination. Although it was largely eclipsed by the rise of modern science from about the middle of the nineteenth century onwards, the sensuous expression of the Romantics was not anti-science. On the contrary they have had a pervasive influence right up to the present day.

From a modern perspective, one of the strangest developments of the Romantic era was the school known as the *Naturphilosophie* or "Nature-philosophy". The Nature-philosophers knew that Newtonian physics must be wrong, because humans have feelings, consciousness and volition which could not be explained by scientific means. Their solution was to permeate the spiritual dimension through everything, thus unifying the Universe instead of perpetuating the division between humans and Creation. Naturphilosophie was inspirational. Goethe and Georg Forster in their own ways sought to understand humanities relationship to nature. Influential in the development of Goethe's ideas was his exposure to eastern traditions, especially Indian holism.....

Both Goethe and Forster deeply influenced Humboldt the father of biogeography and ecology. Through Humboldt we came to see the world as a single inter-connected living system. His influence has been huge including such luminaries as Darwin and Thoreau. . Darwin said he would not have embarked on the Beagle if he had not been influenced by Humboldt. It is not hardtop see Humboldt in Lovelock's idea of Gaia.

The romantic poets were writing in the context of the beginnings of industrialization and a world which seemed more and more mechanistic, and in their works something more sophisticated and much more important was created, representing an emotional relationship with nature. For Wordsworth "Nature" became a medium through which the most profound, even religious thoughts, were conveyed. In his long autobiographical poem, published after his death called Prelude he writes of his childhood

".....even then I felt
Gleams like the flashing of a shield; - the earth
And common face of Nature spake to me
Rememberable things."

Nature could speak through the medium of a "rugged" landscape, an individual primrose growing from a rock-face, or the wind blowing through some trees.

".......or haply, at noon-day,
While in a grove I walk, whose lofty trees,
Laden with summer's thickest foliage, rock
In a strong wind, some working of the spirit,
Some inward agitations thence are brought."

Wild nature was idolized not idealized. As Pope put it in his poem on landscape design, "Of Taste", published in 1731, "Consult the genius of the place in all". Landscaped painters like Friedrich and Turner turned from representing the idyll of a tidy managed countryside to pictures of raw nature. Nature was sanctified again. The pictures of Friedrich are particularly telling, combining as they do a Christian symbolism with wild nature, in effect representing sacred groves.

The transcendentalist movement, particularly in North

America, was influenced by Lorenz Oken, and had its finest flowering in the writings of Henry David Thoreau (1817-1862), Ralph Waldo Emerson (1803-1882), Emily Dickinson (1830-1886), Walt Whitman (1819-1892) and John Muir (1838-1914). It was John Muir who emphasized the value of wildness.

"Walk away quietly in any direction and taste the freedom of the mountaineer. Camp out among the grasses and gentians of glacial meadows, in craggy garden nooks full of nature's darlings. Climb the mountains and get their good tidings, Natures peace flow into you as sunshine flows into trees. The winds will blow their own freshness into you and the storms their energy, while cares will drip off like autumn leaves. As age comes on, one source of enjoyment after another is closed, but nature's sources never fail."

Thoreau in particular has had a lasting impact and has been the inspiration of thousands of aspiring naturalists, particularly in North America but, as a champion for the conservation of nature, there could have been no finer example than John Muir. In Naturphilosophie, the unstable condition of the Universe is a symptom of it being alive. The interplay of two opposites could create an entirely new entity. In recent times this phenomenon is often described as "emergent phenomena".

"There is not a "fragment" in all nature, for every relative fragment of one thing is a full harmonious unit in itself."
"When we try to pick out anything by itself, we find it hitched to everything else in the Universe."

East to Eden

As well idolizing wild nature the Romantic movement turned to the tropics, the location of Eden, but more than that of wild and seemingly untamed nature in which humans lived in a more natural way. The expulsion of humans from Eden and from a harmonious natural world is a very powerful myth. It colors our attitudes to plants and the environment especially to the tropical forest that is supposed to represent it and the tropics seemed to provide a route to return to Eden. Artists like Gauguin and writers like Robert Louis Stevenson were attracted.

European expansion exposed them to vast new lands. Tropical islands seemed to provide new Edens. One of Columbus' motivations was to sail to Eden.

As Grove has very persuasively argued the image, the idea of desert islands (desert meaning uninhabited and natural, not empty) and the bounty of tropical islands as representations of Eden has been a very powerful one. Islands especially tropical islands, their verdant luxury were part of the collective human imagination. Islands because of their isolation provide a stage that brings into sharp focus the human condition. Dante sets purgatory on an Island in the south and Moore set Utopia on an island too.

In a chronicle of Valentine Ferdinand (1495-1561) there is the tale of the English nobleman Sir Robert Machin , on route to Portugal blown off course to landfall on one of the Madeira Islands Porto Santo and was marooned with his page and wife there. Escaping eventually on a raft he reached the shore of Africa and was captured by the Moors and hence news of his discovery came to the world.

Perhaps this story influenced Shakespeare's The Tempest, but he wrote it between 1610-1611at a time when the tales of survivors of the shipwreck on Bermuda (Bermoothes) of the ship "Sea-Adventure", on route to the English colony of Virginia, were fascinating London. Shakespeare seized on the lyrical symbolic potential of the islands , and used it to explore themes such as the wild and untamed (Caliban) against the rational and understood (Prospero). Andrew Marvell's poem entitled Bermuda (Bermoothes) took up this image and developed the expression of the romantic and religious idea of tropical islands, " an isle so long unknown, and yet far kinder than our own, where there is eternal spring, fruit of various kinds and cedars of Lebanon."

Daniel Defoe has Robinson Crusoe exploring his island

"I descended a little on the Side of that delicious Vale, surveying it with a secret Kind of Pleasure, (though mixed with my other afflicting Thoughts) to think that this was all my own, that I was King and Lord of all this Country indefeasibly, and had a Right of Possession; and if I could convey it, I might have it in Inheritance, as completely as any Lord of a Manor in England. I saw here Abundance of Cocoa Trees, Orange, and Lemmon, and Citron

*Trees; but all wild, and very few bearing any Fruit, at least not then:
However, the green Limes that I gathered, were not only pleasant to
eat, but very wholesome; and I mixed their Juice afterwards with
Water, which made it very wholesome, and very cool, and
refreshing. ....I contemplated with great Pleasure the Fruitfulness
of that Valley, and the Pleasantness of the Situation.* "

The island becomes a metaphor for the world and human
lordship. In "Coral Island" and "Swiss Family Robinson" it is a
setting to propound the myth of western moral superiority and
colonial "improvement". But it also becomes the stage on which to
present our anxieties.

The tropical forest, and its inhabitants, noble savages, are also
used as a metaphor for human nature, as if in their wildness represent
the pre-lapsarian natural state of humankind. The tropical idyll
attracts artists like Gauguin and writers like Robert Louis Stevenson.
For Joseph Conrad it was of course the Heart of Darkness, a "dark-
faced pensive forest", "an immensity of trees", or rather the physical
manifestation of Kurtz's dark heart. In this microcosm of the
tropical island the nature of being human and the relationship of
humans to nature, including their own nature is brought sharply into
focus. In the twentieth century, in some of the most moving lines in
English literature William Golding describes the destruction of the
Edenic island in "Lord of the Flies" "For a moment (Ralph) had a
fleeting picture of the strange glamour that once invested the
beaches. But the island was scorched like dead wood .... Ralph wept
for the end of innocence, the darkness of man's heart, and the fall
through the air of the true, wise friend called Piggy". It is Piggy who
is the guardian of rationality in the book. Lord of the Flies was
written after two world wars and the pessimism is very bleak but
later in his novel the Spire, and after the failure of that stone object
to bring about transcendence, Golding turns to different image from
nature as an image of resolution and peace, the image of the bough
of an apple-tree in blossom

*"a cloud of angels flashing in the sunlight , they were pink and
gold and white: they were uttering this sweet scent for joy of the
light and the air."*

William Golding was a close friend of James Lovelock who expanded the island myth to the whole earth with his Gaia hypothesis, ideas developed in the 1960s and published in 1979, treating mother earth as a single organism in which all parts, notably its botanical component interact and moderate changes.

The destruction of Eden

The exploitation and degradation of the tropical lands brought into focus the way humans could have a profound impact on nature by profoundly altering natural ecosystems. The timing of the human caused environmental changes vary over a long time-scale, though only a blink in the history of the earth, from tens of thousands of years ago up to the present.

It is the use of fire which uniquely defines the human species and it was by the use of fire that humans, even pre-agricultural humans, first made a profound impact on natural ecosystems. The use of fire separated humans from the animals: gathered around the campfire they were protected from wild beasts. The use of fire also profoundly changed their economic relationship with nature. Cooking fires made a wider range of plants palatable to eat as root crops could be softened by roasting or boiling and toxic constituents destroyed and it was through the use of fire that humans could manipulate nature, changing it to their own purposes. This was first experienced in arid or semi-arid habitats. In Australia perhaps from 40,000 years ago. …….

Fire was used to corral game in order to hunt it more efficiently. Fire destroyed trees and opened areas for grassland that attracted game. Fire created a more diverse landscape but it was also through its agency that humans may have caused the major extinctions of the megafauna of large herbivores that occurred in Europe and North and South America if not Australia, though rapid, rapidly fluctuating climate change at the end of the last glacial period may have made mega-faunal populations especially vulnerable. Humans are most closely implicated in the extinctions in North America , from 12,000 years ago, in Madagascar about 1500 years ago, also during a period of climate change, and in New Zealand from 900 years ago, each at a time when humans arrived in numbers into ecosystems that had previously not experienced them.

However it was the spread of agriculture and pastoralism that has had the most profound effect on natural ecosystems. The spread of settlements around the Mediterranean basin and the Near East profoundly changed the environment. Fire and pastoral use transformed the vegetation of the vegetation in the Mediterranean basin converting oak, chestnut and pine woodlands to the characteristic mosaic of steppe, spiny scrub (maquis and garigue), orchards and fields. As agriculture and settlements spread and grew in size large areas became deforested. Timber was the primary constructional material and wood was required to bake bricks and produce mortar.

An important consequence of deforestation and agriculture was extensive soil erosion. In many places terracing was established to trap the remaining soil but over vast areas, especially where there were thin fragile soils over limestone karst landscapes, exposed limestone landscapes were created. As fields were abandoned grazing by sheep and goats prevented the regeneration of the former woodlands. There were broader concerns too about how human activity was changing the world; as the environment became increasingly degraded, the past, an idyll of verdant bowers, was increasingly romanticized. There is some evidence that environmental degradation in Anatolia led to the decline of the earliest communities. In the third century B.C.E. Theophrastus voiced concerns about the loss of forests in Greece. Astonishingly he also made the connection between the loss of woodlands and the drying climate.

It is astonishing to think how rapidly and how profoundly the landscape has been changed by agricultural human beings, starting with Neolithic culture perhaps a little over 10,000 years ago, but mainly in the last two centuries. Within a few thousand years huge areas of the world's vegetation has become dominated by the influence of human beings, creating patchworks of field and grassland and creating new plant communities and even new kinds of plants. There are those who see no dangers in our relentless exploitation of nature. The clearing of tropical forest is only the latest phase of our exploitation the world. It is not just western market-led forces and globalization that are responsible. Even before the extinction of the megafauna of the Americas, Eurasia, Australia and New Zealand humanity has profoundly influenced natural ecosystem. For thousands of years Paleolithic hunters had used fire as a means of managing game, corralling it so it could be slaughtered and creating grasslands where it congregated. Forests were exploited to extinction. In the Mediterranean basin and other similar lands new kinds of vegetation arose on soils depleted and eroded as a result of agriculture and pastoralism, though as Rackham has pointed out catastrophic change can arise through natural processes in a changing climate. Nevertheless the hand of humans is heavy in transforming the landscape by draining wetlands or clearing forests. The wildwood has been fragmented.

By 1300 much of central England was an open sea of arable fields. There was intense pressure on other kinds of land. Nature, areas of "waste" declined. The wood became an important possession so that its boundary was marked by a wood-bank and ditch and protected by law. The use of woods as particular kind of pasture especially for pigs but also for deer and cattle was well established. Pannage, the right to fatten pigs in the woodland was one way in which the extent of the woodland was measured, as a wood for so many swine. In pasture woods the trees were pollarded so that the tender branches which provided poles and rods were protected from grazing by growing on a short trunk above grazing height and cut every 10-20 years. Gradually pasture woods had a tendency to be converted to open grassland as woodland regeneration was prevented. A sophisticated management of the resource was maintained by tradition backed up by law. Grazing rights and wood rights were jealously protected. By the thirteenth century much of the lowland wood pasture had disappeared because of too much grazing. It was replaced by open common land. It is only relatively recently in the last 150 years as grazing pressure declined that many of these commons became partially wooded again. In a few areas there was too little grazing and secondary woodland arose. More woodland was planted or arose in hedges around fields. The English countryside was born, not quite artificial, but not wild nature by any means, and quite beautiful in its various ways and more varied, probably more biologically diverse than the "pristine" but probably rather monotonous wildwood.

In some places the environmental changes brought about by humans have been catastrophic leading to the collapse of the human culture. Easter Island, Rap Nui, provides a salutary example.

Colonized between 400-600 C.E. Slash and burn agriculture and population growth led to the destruction of the native forests of Easter Island palm (*Paschalococcus disperta*), toromiro (*Sophora toromiro*) and hau tree (*Hibiscus tilaceus*) that provided fibers for rope. The palm, that may have been used to make rollers for transporting statues or to construct canoes, or a source of edible palm hearts, became extinct about 500 years ago. By 1917 there was only one surviving toromiro tree and that was exterminated by grazing in 1962. Plants grown from seed survived in botanic gardens. Another species *Triumfetta semitriloba*, a woody shrub that provided fibers for textiles was reduced to a four known plants by 1988. One consequence of the loss of woodland was extensive soil erosion leading to a decline in the agricultural productivity and famine, cannibalism and tribal warfare that accompanied a reduction of the population by 1700, to a quarter or less than it had been at its peak.

Deforestation and environmental decline is implicated in the collapse of several civilizations. The ability of forest to attract rain and to trap moisture, and prevent run-off pro-longing its residence in the soil is well known. For example in the region of the Maya a sequence of drier periods about 810, 860, and 910 C.E. within an prolonged climatic drying, were experienced as catastrophic droughts, leading to a collapse of their society from a peak population of about 15 million to a small fraction of that number.

Christopher Columbus according to Grove was one of the first people to make explicit in post-classical times the connection between the deforestation he had observed in the Canaries and Madeira islands, contrasting their severely reduced incidence of mist and rain with the luxuriant rains of the West Indian Islands, with their as yet undamaged forests.

It is in this context that we must evaluate western concerns about the loss of tropical forests. For some the concerns of the western green movement, the campaigns to preserve surviving "pristine" wild nature are just another example of the west trying to assert its cultural and commercial hegemony though few areas of tropical vegetation pristine. They have experienced millennia of utilization by humans. Likewise, it poorly behooves nations that have eliminated most of their own natural vegetation to preach to others about what to do. By the 11th century, as recorded in the

Domesday Book, In the Domesday Book only 3.4% of the England is recorded as woodland, almost entirely as either coppice woodland or as wood pasture.

The west should just butt out, it is up to native people to decide what to do with their land, their nature their plants. But this is to grossly oversimplify the argument. It assumes that some native peoples are not as concerned about the loss of their forests as are green campaigners. It ignores the fact that in a globalized world with a huge and burgeoning population the tropical forests are everyone's backyards.

The most diverse and ancient vegetation, the tropical forests managed to escape very much destruction until the last century but earlier it was the tropical island paradises, important as places in which to re-provision and re-water ships which first experienced the rapid environmental degradation that is so closely linked to western commercial hegemony and globalization. Within the lifetimes of individual observers steep environmental decline took place.

The first to experience the catastrophe of western cultural hegemony were the Fortunate Islands. The islands were known to Carthaginian and Romans perhaps even the Greeks, but did not colonize them. On the Canaries, but apparently not Madeira, there was a stone-age indigenous population probably from North Africa but with blue eyes and long fair hair, the Guanches, whose life-style did not damage the forests. First western colonization came on Lanzarote in the Canaries in 1312 though in 20 years the colonists had been driven off by the natives. On the Canaries a fight back by the Guanches, in a battle in 1494 they killed 2,000 Spanish, only temporarily delayed the progress of colonization. Weakened by plague they were hunted to extinction or sold off into slavery. Normans, Italians, Catalans and Portuguese all competed for the islands. The Portuguese and Spanish explorers and colonists, coming from their own lands which had long been cleared of trees were struck by their wooded appearance. On a Genoese map of 1351 Madeira is figured and titled Isla di Lolegname and later Ilha da Madeira both meaning "wooded isle". Portuguese colonization of the Madeira Islands came in the first half of the 15th century. One of the first thing the colonists did was to clear land for cultivation by firing the woodland. The Portuguese were driven back by the fire for two days to shelter on their ship. Firing of the woodland continued for seven years. Perennial streams dried up and levada, narrow irrigation channels carrying water from the hills were constructed to irrigate the sugar plantations, vineyards and cereal fields, and terraces built to conserved the soil.

Exploitation and loss of habits has either caused the extinction or brought to the edge of extinction many plants. For example California has 20% of plant species at risk. The threats are several. Wild tree species with particularly favored timber have been particular endangered by over-logging. The most important danger is habitat loss or degradation either for urban development or for agriculture and pasture. Climate change, in part brought on by human activity may make extinct small populations at the limits of the current tolerance. One of the most damaging results of human activities has been the introduction of plants to new areas. Humans have introduced exotic species everywhere they have colonized and in some places these have become a significant threat to the native flora. Sometimes concern about introduced plants is exaggerated. The flora of Britain is anyway an entirely immigrant one with very low levels of endemism. However in the wet woodlands of western Scotland that are a hotspot for bryophyte diversity *Rhododendron ponticum*, once native but re-introduced as a garden plant, poses a significant threat to biodiversity.

Regions which have been isolated for a long time and have a high proportion of endemic plants are particularly threatened by introductions. For example some island floras, where there is a high degree of endemism, either ancient and relic, or peculiar and the result of relatively recent adaptive radiation, have been particularly susceptible to introduced species. For example Hawaii for example has 1,200 endemic plant species, 90% of the total native flora. A third of these are rare, and about 150 have less than 50 individuals. In New Zealand 88 species are specifically listed as pest plants including *Clematis vitalba* (old man's beard), wild ginger, purple pampas grass (*Cortaderia jubata*) and *Pinus contorta.* New Zealand's worst weed the gorse *Ulex europaeus, has had* millions of dollars and much effort expended on it to try and control its spread, but paradoxically if left it acts as a nurse plant for native bush protecting native plants from grazing while they establish. In many areas it is aquatic and riverine environments that are especially vulnerable because of the rapidity that the exotic weed can spread along waterways.

A different argument is that extinction does not matter, indeed is a natural phenomenon. Patterns of changing plant distribution, especially during the dramatic climatic fluctuations of the Ice Ages, emphasize the remarkable ability for plant communities to adapt and respond to changing conditions. However pollution like the effect of acid rain has concentrated our attention on the environmental damage that industrialization has caused in less than 200 years. However, all is not lost. We can be sure the plants will respond to the changes, including the climatic change that now seems inevitable. They have responded in innumerable ways in the past. Their present diversity is a record of that evolution. And we can be sure that even if mankind pollutes himself off the biosphere, the plants will remain, evolving anew and making the planet habitable again. Even global warming and high carbon dioxide levels is favoring the spread of trees.

The most important and valuable form of conservation has been the establishment of in situ nature reserves and wild life parks. The number of protected areas has grown rapidly but there are signs that this period of growth is ending because the size of the areas becoming designated as protected is declining. The majority are < 100 ha. According to a UN sponsored survey 8% of the remaining world's forests are in protected areas but unfortunately this does not always necessarily protect some of them from logging or encroachment for other uses especially in south-east Asia where human populations are growing rapidly. And other kinds of vegetation are relatively unprotected. Concerns about the fate of the world's plants has encouraged the establishment of seed banks to preserve samples of seed especially for crop plants and their relatives. The Millennium Seed Bank Project at the Royal Botanic Gardens Kew aims to collect and conserve 10% of the world's flora. Plants from dry-lands have been targeted especially in part because they are regarded as especially vulnerable from climate change and over-grazing but also because seeds from the wet tropics often prove to be rather difficult to store.

Conservationist have resorted to economic arguments in order to promote conservation. Ecotourism has been seen as the savior of our biodiversity. But this is a dangerous mistake because it exposes our surviving natural plantscapes to the vagaries of the marketplace.

Rather we must promote the conservation of biodiversity, of which we are part, as one of the fundamental human values. Plants have created the terrestrial environment we live in and make it habitable for us. All the myriad kinds of terrestrial animals rely upon them too. Each plant species represents a unique point, the result of a unique evolutionary dance, in space and time. Isn't that reason enough?

We cannot return to Eden? It is a myth, a powerful and useful one, but still a myth. Neither can we ignore the huge benefits that the exploitation of plants has brought. We must create Eden, and based on a new set of values, forge a new relationship with nature and the plants that shelter it. But we must not retreat into what the Pope Benedict XVI called vague mysticism. Tree-hugging is nice but daft. We can take our cue from the Bible.... God gave humans dominion over all of the earth and told us to name all its variety – it is time for us to accept that guardianship and re-establish our sacred groves.

"You can drive out nature with a pitchfork, but she keeps on coming back." Horace

"The earth laughs at him who calls a place his own." Hindustani proverb

---